Airman Knowledge Testing Supplement for Sport Pilot, Recreational Pilot, and Private Pilot

2013

This publication combines publications formerly known as:
"Computer Testing Supplement for Recreational Pilot and Private Pilot" and
"Computer Testing Supplement for Sport Pilot, Sport Pilot Instructor,
and Sport Pilot Examiner."

U.S. Department of Transportation
FEDERAL AVIATION ADMINISTRATION
Flight Standards Service

Preface

This Airman Knowledge Testing Supplement is designed by the Federal Aviation Administration (FAA) Flight Standards Service. It is intended for use by Airman Knowledge Testing (AKT) Organization Designation Authorization (ODA) Holders and other entities approved and/or authorized to administer airman knowledge tests on behalf of the FAA in the following knowledge areas:

Sport Pilot–Airplane (ASEL and ASES)
Sport Pilot–Gyroplane
Sport Pilot–Glider
Sport Pilot–Airship (LTA)
Sport Pilot–Balloon (LTA)
Sport Pilot–Weight-Shift Control (WSCS and WSCL)
Sport Pilot–Powered Parachute (PPL and PPS)

Recreational Pilot–Airplane (RPA)
Recreational Pilot–Rotorcraft/Helicopter (RPH)
Recreational Pilot–Rotorcraft/Gyroplane (RPG)

Private Pilot–Airplane/Recreational Pilot–Transition (PAT)
Private Pilot–Helicopter/Recreational Pilot–Transition (PHT)
Private Pilot–Gyroplane/Recreational Pilot–Transition (POT)
Private Pilot–Airplane (PAR)
Private Pilot–Rotorcraft/Helicopter (PRH)
Private Pilot–Rotorcraft/Gyroplane (PRO)
Private Pilot–Glider (POL)
Private Pilot–Free Balloon–Hot Air (PBH)
Private Pilot–Free Balloon–Gas (PBG)
Private Pilot–Lighter-Than-Air–Airship (PLA)
Private Pilot–Powered-Parachute (PPP)
Private Pilot–Weight-shift Control (PWS)

Comments regarding this supplement, or any AFS-630 publication, should be sent in email form to the following address:

AFS630comments@faa.gov

Contents

Appendix 1

SECTIONAL AERONAUTICAL CHART

SCALE 1:500,000

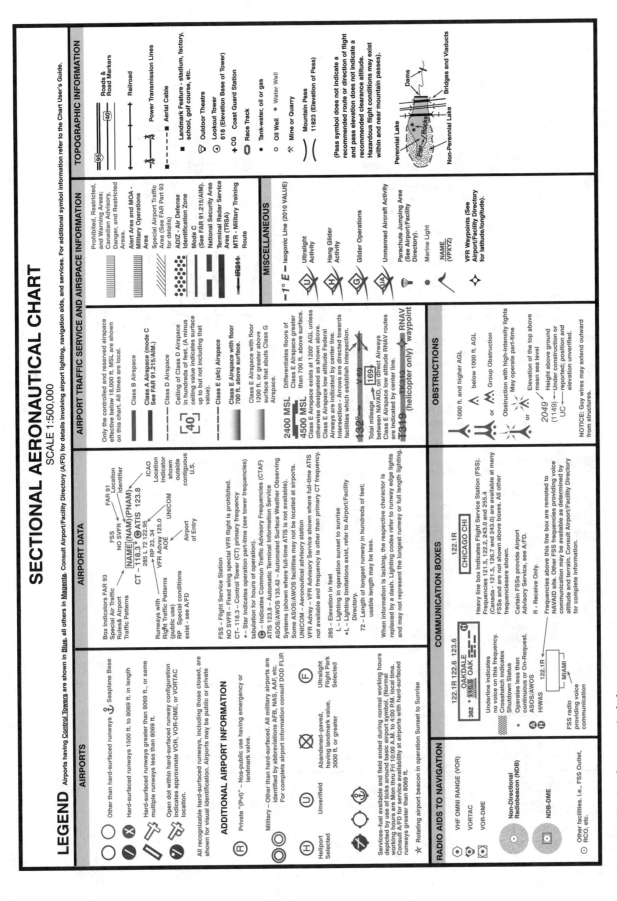

Legend 1. *Sectional aeronautical chart.*

DIRECTORY LEGEND
SAMPLE

CITY NAME
AIRPORT NAME (ALTERNATE NAME) (LTS) (KLTS) CIV/MIL 3 N UTC–6(–5DT) N34°41.93' W99°20.20' JACKSONVILLE
200 B S4 **FUEL** 100 OX 1 TPA—1000(800) AOE Class IV, ARFF Index A NOTAM FILE ORL Not insp. **COPTER**
H–4G, L–19C
IAP, DIAP, AD

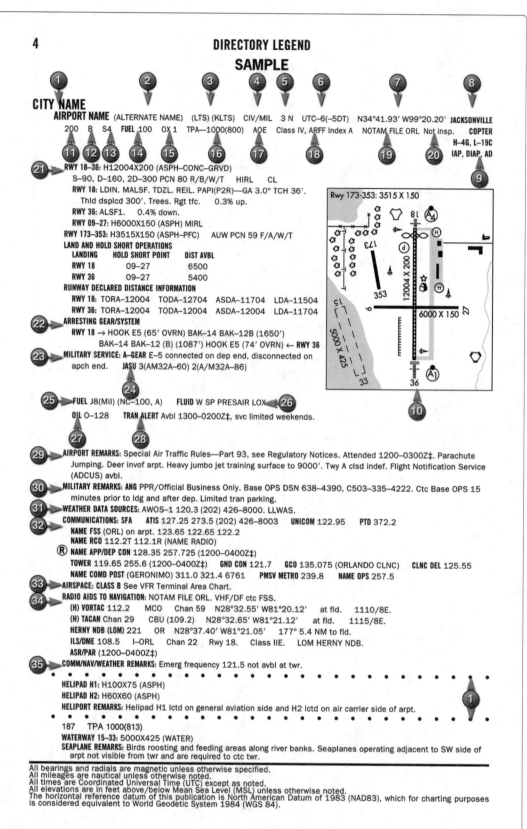

RWY 18–36: H12004X200 (ASPH–CONC–GRVD)
S–90, D–160, 2D–300 PCN 80 R/B/W/T HIRL CL
RWY 18: LDIN. MALSF. TDZL. REIL. PAPI(P2R)—GA 3.0° TCH 36'.
Thld dsplcd 300'. Trees. Rgt tfc. 0.3% up.
RWY 36: ALSF1. 0.4% down.
RWY 09–27: H6000X150 (ASPH) MIRL
RWY 173–353: H3515X150 (ASPH–PFC) AUW PCN 59 F/A/W/T
LAND AND HOLD SHORT OPERATIONS

LANDING	HOLD SHORT POINT	DIST AVBL
RWY 18	09–27	6500
RWY 36	09–27	5400

RUNWAY DECLARED DISTANCE INFORMATION

	TORA	TODA	ASDA	LDA
RWY 18:	TORA–12004	TODA–12704	ASDA–11704	LDA–11504
RWY 36:	TORA–12004	TODA–12004	ASDA–12004	LDA–11704

ARRESTING GEAR/SYSTEM
RWY 18 → HOOK E5 (65' OVRN) BAK–14 BAK–12B (1650')
BAK–14 BAK–12 (B) (1087') HOOK E5 (74' OVRN) ← RWY 36
MILITARY SERVICE: A–GEAR E–5 connected on dep end, disconnected on
apch end. **JASU** 3(AM32A–60) 2(A/M32A–86)

FUEL J8(Mil) (NC–100, A) **FLUID** W SP PRESAIR LOX
OIL O–128 **TRAN ALERT** Avbl 1300–0200Z‡, svc limited weekends.

AIRPORT REMARKS: Special Air Traffic Rules—Part 93, see Regulatory Notices. Attended 1200–0300Z‡. Parachute
Jumping. Deer invof arpt. Heavy jumbo jet training surface to 9000'. Twy A clsd indef. Flight Notification Service
(ADCUS) avbl.
MILITARY REMARKS: ANG PPR/Official Business Only. Base OPS DSN 638–4390, C503–335–4222. Ctc Base OPS 15
minutes prior to ldg and after dep. Limited tran parking.
WEATHER DATA SOURCES: AWOS–1 120.3 (202) 426–8000. LLWAS.
COMMUNICATIONS: SFA ATIS 127.25 273.5 (202) 426–8003 **UNICOM** 122.95 **PTD** 372.2
NAME FSS (ORL) on arpt. 123.65 122.65 122.2
NAME RCO 112.2T 112.1R (NAME RADIO)
Ⓡ **NAME APP/DEP CON** 128.35 257.725 (1200–0400Z‡)
TOWER 119.65 255.6 (1200–0400Z‡) **GND CON** 121.7 **GCO** 135.075 (ORLANDO CLNC) **CLNC DEL** 125.55
NAME COMD POST (GERONIMO) 311.0 321.4 6761 **PMSV METRO** 239.8 **NAME OPS** 257.5
AIRSPACE: CLASS B See VFR Terminal Area Chart.
RADIO AIDS TO NAVIGATION: NOTAM FILE ORL. VHF/DF ctc FSS.
(H) VORTAC 112.2 MCO Chan 59 N28°32.55' W81°20.12' at fld. 1110/8E.
(H) TACAN Chan 29 CBU (109.2) N28°32.65' W81°21.12' at fld. 1115/8E.
HERNY NDB (LOM) 221 OR N28°37.40' W81°21.05' 177° 5.4 NM to fld.
ILS/DME 108.5 I–ORL Chan 22 Rwy 18. Class IIE. LOM HERNY NDB.
ASR/PAR (1200–0400Z‡)
COMM/NAV/WEATHER REMARKS: Emerg frequency 121.5 not avbl at twr.

· ·
HELIPAD H1: H100X75 (ASPH)
HELIPAD H2: H60X60 (ASPH)
HELIPORT REMARKS: Helipad H1 lctd on general aviation side and H2 lctd on air carrier side of arpt.
· ·
187 TPA 1000(813)
WATERWAY 15–33: 5000X425 (WATER)
SEAPLANE REMARKS: Birds roosting and feeding areas along river banks. Seaplanes operating adjacent to SW side of
arpt not visible from twr and are required to ctc twr.

All bearings and radials are magnetic unless otherwise specified.
All mileages are nautical unless otherwise noted.
All times are Coordinated Universal Time (UTC) except as noted.
All elevations are in feet above/below Mean Sea Level (MSL) unless otherwise noted.
The horizontal reference datum of this publication is North American Datum of 1983 (NAD83), which for charting purposes
is considered equivalent to World Geodetic System 1984 (WGS 84).

NE, 09 FEB 2012 to 05 APR 2012

Legend 2. *Airport/facility directory.*

10

SKETCH LEGEND

RUNWAYS/LANDING AREAS

Hard Surfaced . . . .

Metal Surface . . . .

Sod, Gravel, etc. . . . .

Light Plane, ...
Ski Landing Area or Water

Under Construction . . . . .

Closed . . . .

Helicopter Landings Area . . . . (H)

Displaced Threshold . . . .

Taxiway, Apron and Stopways . .

RADIO AIDS TO NAVIGATION

VORTAC . . .   VOR . . . .

VOR/DME . .   NDB . . . .

TACAN . . . .   NDB/DME . . . .

MISCELLANEOUS AERONAUTICAL FEATURES

Airport Beacon . . . .   ☆ ✪

Wind Cone . . . .

Landing Tee . . . .

Tetrahedron . . . .

Control Tower . . . .   or TWR

When control tower and rotating beacon
are co-located beacon symbol will be
used and further identified as TWR.

MISCELLANEOUS BASE AND CULTURAL FEATURES

Buildings . . . .

Power Lines . . . .   ─T──T─

Fence . . . .   ×─×─×─×─×─×

Towers . . . .

Tanks . . . .

Oil Well . . . .

Smoke Stack . . . .

Obstruction . . . .   5812 Λ

Controlling Obstruction . . . .   +5812

Trees . . . .

Populated Places . . . .

Cuts and Fills . . . . . . Cut   Fill

Cliffs and Depressions . .

Ditch . . . .

Hill . . . .

APPROACH LIGHTING SYSTEMS

A dot '•' portrayed with approach lighting
letter identifier indicates sequenced flashing
lights (F) installed with the approach lighting
system e.g. (A₁) Negative symbology, e.g., (A₁)
(V) indicates Pilot Controlled Lighting (PCL).

Runway Centerline Lighting . . . . .

(A) Approach Lighting System ALSF-2 . .

(A₁) Approach Lighting System ALSF-1 . .

(A₂) Short Approach Lighting System
    SALS/SALSF . . . .

(A₃) Simplified Short Approach Lighting
    System (SSALR) with RAIL . . . . .

(A₄) Medium Intensity Approach Lighting System
    (MALS and MALSF)/(SSALS

(A₄) and SSALF) . . . .

(A₅) Medium Intensity Approach Lighting
    System (MALSR) and RAIL . . . . . .

    Omnidirectional Approach
    Lighting System (ODALS) . . . .

(D) Navy Parallel Row and Cross Bar . . .

(+) Air Force Overrun . . . .

(V) Visual Approach Slope Indicator with
    Standard Threshold Clearance provided

(V₂) Pulsating Visual Approach Slope Indicator
    (PVASI)

(V₃) Visual Approach Slope Indicator with a
    threshold crossing height to accomodate
    long bodied or jumbo aircraft

(V₄) Tri-color Visual Approach Slope Indicator
    (TRCV)

(V₅) Approach Path Alignment Panel (APAP)

(P) Precision Approach Path Indicator (PAPI)

NE, 09 FEB 2012 to 05 APR 2012

Legend 3. *Airport/facility directory.*

LEGEND

This directory is a listing of data on record with the FAA on all open to the public airports, military facilities and selected private use facilities specifically requested by the Department of Defense (DoD) for which a DoD Instrument Approach Procedure has been published in the U.S. Terminal Procedures Publication. Additionally this listing contains data for associated terminal control facilities, air route traffic control centers, and radio aids to navigation within the conterminous United States, Puerto Rico and the Virgin Islands. Joint civil/military and civil airports are listed alphabetically by state, associated city and airport name and cross-referenced by airport name. Military facilities are listed alphabetically by state and official airport name and cross-referenced by associated city name. Navaids, flight service stations and remote communication outlets that are associated with an airport, but with a different name, are listed alphabetically under their own name, as well as under the airport with which they are associated.

The listing of an open to the public airport in this directory merely indicates the airport operator's willingness to accommodate transient aircraft, and does not represent that the facility conforms with any Federal or local standards, or that it has been approved for use on the part of the general public. Military and private use facilities published in this directory are open to civil pilots only in an emergency or with prior permission. See Special Notice Section, Civil Use of Military Fields.

The information on obstructions is taken from reports submitted to the FAA. Obstruction data has not been verified in all cases. Pilots are cautioned that objects not indicated in this tabulation (or on the airports sketches and/or charts) may exist which can create a hazard to flight operation. Detailed specifics concerning services and facilities tabulated within this directory are contained in the Aeronautical Information Manual, Basic Flight Information and ATC Procedures.

The legend items that follow explain in detail the contents of this Directory and are keyed to the circled numbers on the sample on the preceding pages.

 CITY/AIRPORT NAME

Civil and joint civil/military airports and facilities in this directory are listed alphabetically by state and associated city. Where the city name is different from the airport name the city name will appear on the line above the airport name. Airports with the same associated city name will be listed alphabetically by airport name and will be separated by a dashed rule line. A solid rule line will separate all others. FAA approved helipads and seaplane landing areas associated with a land airport will be separated by a dotted line. Military airports are listed alphabetically by state and official airport name.

 ALTERNATE NAME

Alternate names, if any, will be shown in parentheses.

 LOCATION IDENTIFIER

The location identifier is a three or four character FAA code followed by a four-character ICAO code assigned to airports. ICAO codes will only be published at joint civil/military, and military facilities. If two different military codes are assigned, both codes will be shown with the primary operating agency's code listed first. These identifiers are used by ATC in lieu of the airport name in flight plans, flight strips and other written records and computer operations. Zeros will appear with a slash to differentiate them from the letter "O".

 OPERATING AGENCY

Airports within this directory are classified into two categories, Military/Federal Government and Civil airports open to the general public, plus selected private use airports. The operating agency is shown for military, private use and joint civil/military airports. The operating agency is shown by an abbreviation as listed below. When an organization is a tenant, the abbreviation is enclosed in parenthesis. No classification indicates the airport is open to the general public with no military tenant.

A	US Army	MC	Marine Corps
AFRC	Air Force Reserve Command	N	Navy
AF	US Air Force	NAF	Naval Air Facility
ANG	Air National Guard	NAS	Naval Air Station
AR	US Army Reserve	NASA	National Air and Space Administration
ARNG	US Army National Guard	P	US Civil Airport Wherein Permit Covers
CG	US Coast Guard		Use by Transient Military Aircraft
CIV/MIL	Joint Use Civil/Military	PVT	Private Use Only (Closed to the Public)
DND	Department of National Defense Canada		

 AIRPORT LOCATION

Airport location is expressed as distance and direction from the center of the associated city in nautical miles and cardinal points, e.g., 4 NE.

 TIME CONVERSION

Hours of operation of all facilities are expressed in Coordinated Universal Time (UTC) and shown as "Z" time. The directory indicates the number of hours to be subtracted from UTC to obtain local standard time and local daylight saving time UTC–5(–4DT). The symbol ‡ indicates that during periods of Daylight Saving Time effective hours will be one hour earlier than shown. In those areas where daylight saving time is not observed the (–4DT) and ‡ will not be shown. Daylight saving time is in effect from 0200 local time the second Sunday in March to 0200 local time the first Sunday in November. Canada and all U.S. Conterminous States observe daylight saving time except Arizona and Puerto Rico, and the Virgin Islands. If the state observes daylight saving time and the operating times are other than daylight saving times, the operating hours will include the dates, times and no ‡ symbol will be shown, i.e., April 15–Aug 31 0630–1700Z, Sep 1–Apr 14 0600–1700Z.

NE, 09 FEB 2012 to 05 APR 2012

Legend 4. *Airport/facility directory.*

GEOGRAPHIC POSITION OF AIRPORT—AIRPORT REFERENCE POINT (ARP)

Positions are shown as hemisphere, degrees, minutes and hundredths of a minute and represent the approximate geometric center of all usable runway surfaces.

 CHARTS

Charts refer to the Sectional Chart and Low and High Altitude Enroute Chart and panel on which the airport or facility is located. Helicopter Chart locations will be indicated as COPTER. IFR Gulf of Mexico West and IFR Gulf of Mexico Central will be depicted as GOMW and GOMC.

9 **INSTRUMENT APPROACH PROCEDURES, AIRPORT DIAGRAMS**

IAP indicates an airport for which a prescribed (Public Use) FAA Instrument Approach Procedure has been published. DIAP indicates an airport for which a prescribed DoD Instrument Approach Procedure has been published in the U.S. Terminal Procedures. See the Special Notice Section of this directory, Civil Use of Military Fields and the Aeronautical Information Manual 5–4–5 Instrument Approach Procedure Charts for additional information. AD indicates an airport for which an airport diagram has been published. Airport diagrams are located in the back of each A/FD volume alphabetically by associated city and airport name.

10 **AIRPORT SKETCH**

The airport sketch, when provided, depicts the airport and related topographical information as seen from the air and should be used in conjunction with the text. It is intended as a guide for pilots in VFR conditions. Symbology that is not self-explanatory will be reflected in the sketch legend. The airport sketch will be oriented with True North at the top. Airport sketches will be added incrementally.

11 **ELEVATION**

The highest point of an airport's usable runways measured in feet from mean sea level. When elevation is sea level it will be indicated as "00". When elevation is below sea level a minus "−" sign will precede the figure.

12 **ROTATING LIGHT BEACON**

B indicates rotating beacon is available. Rotating beacons operate sunset to sunrise unless otherwise indicated in the AIRPORT REMARKS or MILITARY REMARKS segment of the airport entry.

13 **SERVICING—CIVIL**

S1: Minor airframe repairs.	S5: Major airframe repairs.
S2: Minor airframe and minor powerplant repairs.	S6: Minor airframe and major powerplant repairs.
S3: Major airframe and minor powerplant repairs.	S7: Major powerplant repairs.
S4: Major airframe and major powerplant repairs.	S8: Minor powerplant repairs.

 FUEL

CODE	FUEL	CODE	FUEL
80	Grade 80 gasoline (Red)	B+	Jet B, Wide-cut, turbine fuel with FS–II*, FP** minus 50° C.
100	Grade 100 gasoline (Green)		
100LL	100LL gasoline (low lead) (Blue)	J4 (JP4)	(JP–4 military specification) FP** minus 58° C.
115	Grade 115 gasoline (115/145 military specification) (Purple)	J5 (JP5)	(JP–5 military specification) Kerosene with FS–11, FP** minus 46°C.
A	Jet A, Kerosene, without FS–II*, FP** minus 40° C.	J8 (JP8)	(JP–8 military specification) Jet A–1, Kerosene with FS–II*, FP** minus 47°C.
A+	Jet A, Kerosene, with FS–II*, FP** minus 40°C.	J8+100	(JP–8 military specification) Jet A–1, Kerosene with FS–II*, FP** minus 47°C, with-fuel additive package that improves thermo stability characteristics of JP-8.
A1	Jet A–1, Kerosene, without FS–II*, FP** minus 47°C.		
A1+	Jet A–1, Kerosene with FS–II*, FP** minus 47° C.	J	(Jet Fuel Type Unknown)
B	Jet B, Wide-cut, turbine fuel without FS–II*, FP** minus 50° C.	MOGAS	Automobile gasoline which is to be used as aircraft fuel.

*(Fuel System Icing Inhibitor)

**(Freeze Point)

NOTE: Certain automobile gasoline may be used in specific aircraft engines if a FAA supplemental type certificate has been obtained. Automobile gasoline, which is to be used in aircraft engines, will be identified as "MOGAS", however, the grade/type and other octane rating will not be published.

Data shown on fuel availability represents the most recent information the publisher has been able to acquire. Because of a variety of factors, the fuel listed may not always be obtainable by transient civil pilots. Confirmation of availability of fuel should be made directly with fuel suppliers at locations where refueling is planned.

15 **OXYGEN—CIVIL**

OX 1 High Pressure	OX 3 High Pressure—Replacement Bottles
OX 2 Low Pressure	OX 4 Low Pressure—Replacement Bottles

16 **TRAFFIC PATTERN ALTITUDE**

Traffic Pattern Altitude (TPA)—The first figure shown is TPA above mean sea level. The second figure in parentheses is TPA above airport elevation. Multiple TPA shall be shown as "TPA—See Remarks" and detailed information shall be shown in the Airport or Military Remarks Section. Traffic pattern data for USAF bases, USN facilities, and U.S. Army airports (including those on which ACC or U.S. Army is a tenant) that deviate from standard pattern altitudes shall be shown in Military Remarks.

NE, 09 FEB 2012 to 05 APR 2012

Legend 5. *Airport/facility directory.*

AIRPORT OF ENTRY, LANDING RIGHTS, AND CUSTOMS USER FEE AIRPORTS

U.S. CUSTOMS USER FEE AIRPORT—Private Aircraft operators are frequently required to pay the costs associated with customs processing.

AOE—Airport of Entry. A customs Airport of Entry where permission from U.S. Customs is not required to land. However, at least one hour advance notice of arrival is required.

LRA—Landing Rights Airport. Application for permission to land must be submitted in advance to U.S. Customs. At least one hour advance notice of arrival is required.

NOTE: Advance notice of arrival at both an AOE and LRA airport may be included in the flight plan when filed in Canada or Mexico. Where Flight Notification Service (ADCUS) is available the airport remark will indicate this service. This notice will also be treated as an application for permission to land in the case of an LRA. Although advance notice of arrival may be relayed to Customs through Mexico, Canada, and U.S. Communications facilities by flight plan, the aircraft operator is solely responsible for ensuring that Customs receives the notification. (See Customs, Immigration and Naturalization, Public Health and Agriculture Department requirements in the International Flight Information Manual for further details.)

US Customs Air and Sea Ports, Inspectors and Agents

Northeast Sector (New England and Atlantic States—ME to MD)	407–975–1740
Southeast Sector (Atlantic States—DC, WV, VA to FL)	407–975–1780
Central Sector (Interior of the US, including Gulf states—MS, AL, LA)	407–975–1760
Southwest East Sector (OK and eastern TX)	407–975–1840
Southwest West Sector (Western TX, NM and AZ)	407–975–1820
Pacific Sector (WA, OR, CA, HI and AK)	407–975–1800

 CERTIFICATED AIRPORT (14 CFR PART 139)

Airports serving Department of Transportation certified carriers and certified under 14 CFR part 139 are indicated by the Class and the ARFF Index; e.g. Class I, ARFF Index A, which relates to the availability of crash, fire, rescue equipment. Class I airports can have an ARFF Index A through E, depending on the aircraft length and scheduled departures. Class II, III, and IV will always carry an Index A.

14 CFR PART 139 CERTIFICATED AIRPORTS
AIRPORT CLASSIFICATIONS

Type of Air Carrier Operation	Class I	Class II	Class III	Class IV
Scheduled Air Carrier Aircraft with 31 or more passenger seats	X			
Unscheduled Air Carrier Aircraft with 31 or more passengers seats	X	X		X
Scheduled Air Carrier Aircraft with 10 to 30 passenger seats	X	X	X	

14 CFR–PART 139 CERTIFICATED AIRPORTS
INDICES AND AIRCRAFT RESCUE AND FIRE FIGHTING EQUIPMENT REQUIREMENTS

Airport Index	Required No. Vehicles	Aircraft Length	Scheduled Departures	Agent + Water for Foam
A	1	<90′	≥1	500#DC or HALON 1211 or 450#DC + 100 gal H$_2$O
B	1 or 2	≥90′, <126′ ------------------- ≥126′, <159′	≥5 ------------ <5	Index A + 1500 gal H$_2$O
C	2 or 3	≥126′, <159′ ------------------- ≥159′, <200′	≥5 ------------ <5	Index A + 3000 gal H$_2$O
D	3	≥159′, <200′ ------------------- >200′	 <5	Index A + 4000 gal H$_2$O
E	3	≥200′	≥5	Index A + 6000 gal H$_2$O

> Greater Than; < Less Than; ≥ Equal or Greater Than; ≤ Equal or Less Than;   H$_2$O–Water; DC–Dry Chemical.

NOTE: The listing of ARFF index does not necessarily assure coverage for non-air carrier operations or at other than prescribed times for air carrier. ARFF Index Ltd.—indicates ARFF coverage may or may not be available, for information contact airport manager prior to flight.

 NOTAM SERVICE

All public use landing areas are provided NOTAM service. A NOTAM FILE identifier is shown for individual langing areas, e.g., "NOTAM FILE BNA". See the AIM, Basic Flight Information and ATC Procedures for a detailed description of NOTAMs.

NE, 09 FEB 2012 to 05 APR 2012

Legend 6. *Airport/facility directory.*

Current NOTAMs are available from flight service stations at 1–800–WX–BRIEF (992–7433) or online through the FAA PilotWeb at https://pilotweb.nas.faa.gov. Military NOTAMs are available using the Defense Internet NOTAM Service (DINS) at https://www.notams.jcs.mil.

Pilots flying to or from airports not available through the FAA PilotWeb or DINS can obtain assistance from Flight Service.

 FAA INSPECTION

All airports not inspected by FAA will be identified by the note: Not insp. This indicates that the airport information has been provided by the owner or operator of the field.

 RUNWAY DATA

Runway information is shown on two lines. That information common to the entire runway is shown on the first line while information concerning the runway ends is shown on the second or following line. Runway direction, surface, length, width, weight bearing capacity, lighting, and slope, when available are shown for each runway. Multiple runways are shown with the longest runway first. Direction, length, width, and lighting are shown for sea-lanes. The full dimensions of helipads are shown, e.g., 50X150. Runway data that requires clarification will be placed in the remarks section.

RUNWAY DESIGNATION

Runways are normally numbered in relation to their magnetic orientation rounded off to the nearest 10 degrees. Parallel runways can be designated L (left)/R (right)/C (center). Runways may be designated as Ultralight or assault strips. Assault strips are shown by magnetic bearing.

RUNWAY DIMENSIONS

Runway length and width are shown in feet. Length shown is runway end to end including displaced thresholds, but excluding those areas designed as overruns.

RUNWAY SURFACE AND LENGTH

Runway lengths prefixed by the letter "H" indicate that the runways are hard surfaced (concrete, asphalt, or part asphalt–concrete). If the runway length is not prefixed, the surface is sod, clay, etc. The runway surface composition is indicated in parentheses after runway length as follows:

(AFSC)—Aggregate friction seal coat	(GRVD)—Grooved	(PSP)—Pierced steel plank
(AMS)—Temporary metal planks coated with nonskid material	(GRVL)—Gravel, or cinders	(RFSC)—Rubberized friction seal coat
	(MATS)—Pierced steel planking, landing mats, membranes	(TURF)—Turf
(ASPH)—Asphalt		(TRTD)—Treated
(CONC)—Concrete	(PEM)—Part concrete, part asphalt	(WC)—Wire combed
(DIRT)—Dirt	(PFC)—Porous friction courses	

RUNWAY WEIGHT BEARING CAPACITY

Runway strength data shown in this publication is derived from available information and is a realistic estimate of capability at an average level of activity. It is not intended as a maximum allowable weight or as an operating limitation. Many airport pavements are capable of supporting limited operations with gross weights in excess of the published figures. Permissible operating weights, insofar as runway strengths are concerned, are a matter of agreement between the owner and user. When desiring to operate into any airport at weights in excess of those published in the publication, users should contact the airport management for permission. Runway strength figures are shown in thousand of pounds, with the last three figures being omitted. Add 000 to figure following S, D, 2S, 2T, AUW, SWL, etc., for gross weight capacity. A blank space following the letter designator is used to indicate the runway can sustain aircraft with this type landing gear, although definite runway weight bearing capacity figures are not available, e.g., S, D. Applicable codes for typical gear configurations with S=Single, D=Dual, T=Triple and Q=Quadruple:

CURRENT	NEW	NEW DESCRIPTION
S	S	Single wheel type landing gear (DC3), (C47), (F15), etc.
D	D	Dual wheel type landing gear (BE1900), (B737), (A319), etc.
T	D	Dual wheel type landing gear (P3, C9).
ST	2S	Two single wheels in tandem type landing gear (C130).
TRT	2T	Two triple wheels in tandem type landing gear (C17), etc.
DT	2D	Two dual wheels in tandem type landing gear (B707), etc.
TT	2D	Two dual wheels in tandem type landing gear (B757, KC135).
SBTT	2D/D1	Two dual wheels in tandem/dual wheel body gear type landing gear (KC10).
None	2D/2D1	Two dual wheels in tandem/two dual wheels in tandem body gear type landing gear (A340–600).
DDT	2D/2D2	Two dual wheels in tandem/two dual wheels in double tandem body gear type landing gear (B747, E4).
TTT	3D	Three dual wheels in tandem type landing gear (B777), etc.
TT	D2	Dual wheel gear two struts per side main gear type landing gear (B52).
TDT	C5	Complex dual wheel and quadruple wheel combination landing gear (C5).

NE, 09 FEB 2012 to 05 APR 2012

Legend 7. *Airport/facility directory.*

AUW—All up weight. Maximum weight bearing capacity for any aircraft irrespective of landing gear configuration.

SWL—Single Wheel Loading. (This includes information submitted in terms of Equivalent Single Wheel Loading (ESWL) and Single Isolated Wheel Loading).

PSI—Pounds per square inch. PSI is the actual figure expressing maximum pounds per square inch runway will support, e.g., (SWL 000/PSI 535).

Omission of weight bearing capacity indicates information unknown.

The ACN/PCN System is the ICAO standard method of reporting pavement strength for pavements with bearing strengths greater than 12,500 pounds. The Pavement Classification Number (PCN) is established by an engineering assessment of the runway. The PCN is for use in conjunction with an Aircraft Classification Number (ACN). Consult the Aircraft Flight Manual, Flight Information Handbook, or other appropriate source for ACN tables or charts. Currently, ACN data may not be available for all aircraft. If an ACN table or chart is available, the ACN can be calculated by taking into account the aircraft weight, the pavement type, and the subgrade category. For runways that have been evaluated under the ACN/PCN system, the PCN will be shown as a five-part code (e.g. PCN 80 R/B/W/T). Details of the coded format are as follows:

(1) The PCN NUMBER—The reported PCN indicates that an aircraft with an ACN equal or less than the reported PCN can operate on the pavement subject to any limitation on the tire pressure.

(2) The type of pavement:
R — Rigid
F — Flexible

(3) The pavement subgrade category:
A — High
B — Medium
C — Low
D — Ultra-low

(4) The maximum tire pressure authorized for the pavement:
W — High, no limit
X — Medium, limited to 217 psi
Y — Low, limited to 145 psi
Z — Very low, limited to 73 psi

(5) Pavement evaluation method:
T — Technical evaluation
U — By experience of aircraft using the pavement

NOTE: Prior permission from the airport controlling authority is required when the ACN of the aircraft exceeds the published PCN or aircraft tire pressure exceeds the published limits.

RUNWAY LIGHTING

Lights are in operation sunset to sunrise. Lighting available by prior arrangement only or operating part of the night and/or pilot controlled lighting with specific operating hours are indicated under airport or military remarks. At USN/USMC facilities lights are available only during airport hours of operation. Since obstructions are usually lighted, obstruction lighting is not included in this code. Unlighted obstructions on or surrounding an airport will be noted in airport or military remarks. Runway lights nonstandard (NSTD) are systems for which the light fixtures are not FAA approved L-800 series: color, intensity, or spacing does not meet FAA standards. Nonstandard runway lights, VASI, or any other system not listed below will be shown in airport remarks or military service. Temporary, emergency or limited runway edge lighting such as flares, smudge pots, lanterns or portable runway lights will also be shown in airport remarks or military service. Types of lighting are shown with the runway or runway end they serve.

NSTD—Light system fails to meet FAA standards.
LIRL—Low Intensity Runway Lights.
MIRL—Medium Intensity Runway Lights.
HIRL—High Intensity Runway Lights.
RAIL—Runway Alignment Indicator Lights.
REIL—Runway End Identifier Lights.
CL—Centerline Lights.
TDZL—Touchdown Zone Lights.
ODALS—Omni Directional Approach Lighting System.
AF OVRN—Air Force Overrun 1000′ Standard
    Approach Lighting System.
LDIN—Lead-In Lighting System.
MALS—Medium Intensity Approach Lighting System.
MALSF—Medium Intensity Approach Lighting System with Sequenced Flashing Lights.
MALSR—Medium Intensity Approach Lighting System with Runway Alignment Indicator Lights.

SALS—Short Approach Lighting System.
SALSF—Short Approach Lighting System with Sequenced Flashing Lights.
SSALS—Simplified Short Approach Lighting System.
SSALF—Simplified Short Approach Lighting System with Sequenced Flashing Lights.
SSALR—Simplified Short Approach Lighting System with Runway Alignment Indicator Lights.
ALSAF—High Intensity Approach Lighting System with Sequenced Flashing Lights.
ALSF1—High Intensity Approach Lighting System with Sequenced Flashing Lights, Category I, Configuration.
ALSF2—High Intensity Approach Lighting System with Sequenced Flashing Lights, Category II, Configuration.
SF—Sequenced Flashing Lights.
OLS—Optical Landing System.
WAVE–OFF.

NOTE: Civil ALSF2 may be operated as SSALR during favorable weather conditions. When runway edge lights are positioned more than 10 feet from the edge of the usable runway surface a remark will be added in the ''Remarks'' portion of the airport entry. This is applicable to Air Force, Air National Guard and Air Force Reserve Bases, and those joint civil/military airfields on which they are tenants.

NE, 09 FEB 2012 to 05 APR 2012

Legend 8. *Airport/facility directory.*

VISUAL GLIDESLOPE INDICATORS

APAP—A system of panels, which may or may not be lighted, used for alignment of approach path.

PNIL	APAP on left side of runway	PNIR	APAP on right side of runway

PAPI—Precision Approach Path Indicator

P2L	2-identical light units placed on left side of runway	P4L	4-identical light units placed on left side of runway
P2R	2-identical light units placed on right side of runway	P4R	4-identical light units placed on right side of runway

PVASI—Pulsating/steady burning visual approach slope indicator, normally a single light unit projecting two colors.

PSIL	PVASI on left side of runway	PSIR	PVASI on right side of runway

SAVASI—Simplified Abbreviated Visual Approach Slope Indicator

S2L	2-box SAVASI on left side of runway	S2R	2-box SAVASI on right side of runway

TRCV—Tri-color visual approach slope indicator, normally a single light unit projecting three colors.

TRIL	TRCV on left side of runway	TRIR	TRCV on right side of runway

VASI—Visual Approach Slope Indicator

V2L	2-box VASI on left side of runway	V6L	6-box VASI on left side of runway
V2R	2-box VASI on right side of runway	V6R	6-box VASI on right side of runway
V4L	4-box VASI on left side of runway	V12	12-box VASI on both sides of runway
V4R	4-box VASI on right side of runway	V16	16-box VASI on both sides of runway

NOTE: Approach slope angle and threshold crossing height will be shown when available; i.e., –GA 3.5° TCH 37′.

PILOT CONTROL OF AIRPORT LIGHTING

Key Mike	Function
7 times within 5 seconds	Highest intensity available
5 times within 5 seconds	Medium or lower intensity (Lower REIL or REIL-Off)
3 times within 5 seconds	Lowest intensity available (Lower REIL or REIL-Off)

Available systems will be indicated in the airport or military remarks, e.g., ACTIVATE HIRL Rwy 07–25, MALSR Rwy 07, and VASI Rwy 07—122.8.

Where the airport is not served by an instrument approach procedure and/or has an independent type system of different specification installed by the airport sponsor, descriptions of the type lights, method of control, and operating frequency will be explained in clear text. See AIM, "Basic Flight Information and ATC Procedures," for detailed description of pilot control of airport lighting.

RUNWAY SLOPE

When available, runway slope data will only be provided for those airports with an approved FAA instrument approach procedure. Runway slope will be shown only when it is 0.3 percent or greater. On runways less than 8000 feet, the direction of the slope up will be indicated, e.g., 0.3% up NW. On runways 8000 feet or greater, the slope will be shown (up or down) on the runway end line, e.g., RWY 13: 0.3% up., RWY 21: Pole. Rgt tfc. 0.4% down.

RUNWAY END DATA

Information pertaining to the runway approach end such as approach lights, touchdown zone lights, runway end identification lights, visual glideslope indicators, displaced thresholds, controlling obstruction, and right hand traffic pattern, will be shown on the specific runway end. "Rgt tfc"—Right traffic indicates right turns should be made on landing and takeoff for specified runway end.

LAND AND HOLD SHORT OPERATIONS (LAHSO)

LAHSO is an acronym for "Land and Hold Short Operations." These operations include landing and holding short of an intersection runway, an intersecting taxiway, or other predetermined points on the runway other than a runway or taxiway. Measured distance represents the available landing distance on the landing runway, in feet.

Specific questions regarding these distances should be referred to the air traffic manager of the facility concerned. The Aeronautical Information Manual contains specific details on hold–short operations and markings.

RUNWAY DECLARED DISTANCE INFORMATION

TORA—Take-off Run Available. The length of runway declared available and suitable for the ground run of an aeroplane take–off.

TODA—Take-off Distance Available. The length of the take–off run available plus the length of the clearway, if provided.

ASDA—Accelerate-Stop Distance Available. The length of the take–off run available plus the length of the stopway, if provided.

LDA—Landing Distance Available. The length of runway which is declared available and suitable for the ground run of an aeroplane landing.

22 ARRESTING GEAR/SYSTEMS

Arresting gear is shown as it is located on the runway. The a–gear distance from the end of the appropriate runway (or into the overrun) is indicated in parentheses. A–Gear which has a bi–direction capability and can be utilized for emergency approach end engagement is indicated by a (B). The direction of engaging device is indicated by an arrow. Up to 15 minutes advance notice may be required for rigging A–Gear for approach and engagement. Airport listing may show availability of other than US Systems. This information is provided for emergency requirements only. Refer to current aircraft operating manuals for specific engagement weight and speed criteria based on aircraft structural restrictions and arresting system limitations.

Following is a list of current systems referenced in this publication identified by both Air Force and Navy terminology:

NE, 09 FEB 2012 to 05 APR 2012

Legend 9. *Airport/facility directory.*

BI–DIRECTIONAL CABLE (B)

TYPE	DESCRIPTION
BAK–9	Rotary friction brake.
BAK–12A	Standard BAK–12 with 950 foot run out, 1–inch cable and 40,000 pound weight setting. Rotary friction brake.
BAK–12B	Extended BAK–12 with 1200 foot run, 1¼ inch Cable and 50,000 pounds weight setting. Rotary friction brake.
E28	Rotary Hydraulic (Water Brake).
M21	Rotary Hydraulic (Water Brake) Mobile.

The following device is used in conjunction with some aircraft arresting systems:

BAK–14	A device that raises a hook cable out of a slot in the runway surface and is remotely positioned for engagement by the tower on request. (In addition to personnel reaction time, the system requires up to five seconds to fully raise the cable.)
H	A device that raises a hook cable out of a slot in the runway surface and is remotely positioned for engagement by the tower on request. (In addition to personnel reaction time, the system requires up to one and one–half seconds to fully raise the cable.)

UNI–DIRECTIONAL CABLE

TYPE	DESCRIPTION
MB60	Textile brake—an emergency one–time use, modular braking system employing the tearing of specially woven textile straps to absorb the kinetic energy.
E5/E5–1/E5–3	Chain Type. At USN/USMC stations E–5 A–GEAR systems are rated, e.g., E–5 RATING–13R–1100 HW (DRY), 31L/R–1200 STD (WET). This rating is a function of the A–GEAR chain weight and length and is used to determine the maximum aircraft engaging speed. A dry rating applies to a stabilized surface (dry or wet) while a wet rating takes into account the amount (if any) of wet overrun that is not capable of withstanding the aircraft weight. These ratings are published under Military Service.

FOREIGN CABLE

TYPE	DESCRIPTION	US EQUIVALENT
44B–3H	Rotary Hydraulic) (Water Brake)	
CHAG	Chain	E–5

UNI–DIRECTIONAL BARRIER

TYPE	DESCRIPTION
MA–1A	Web barrier between stanchions attached to a chain energy absorber.
BAK–15	Web barrier between stanchions attached to an energy absorber (water squeezer, rotary friction, chain). Designed for wing engagement.

NOTE: Landing short of the runway threshold on a runway with a BAK–15 in the underrun is a significant hazard. The barrier in the down position still protrudes several inches above the underrun. Aircraft contact with the barrier short of the runway threshold can cause damage to the barrier and substantial damage to the aircraft.

OTHER

TYPE	DESCRIPTION
EMAS	Engineered Material Arresting System, located beyond the departure end of the runway, consisting of high energy absorbing materials which will crush under the weight of an aircraft.

23 MILITARY SERVICE

Specific military services available at the airport are listed under this general heading. Remarks applicable to any military service are shown in the individual service listing.

24 JET AIRCRAFT STARTING UNITS (JASU)

The numeral preceding the type of unit indicates the number of units available. The absence of the numeral indicates ten or more units available. If the number of units is unknown, the number one will be shown. Absence of JASU designation indicates non–availability.

The following is a list of current JASU systems referenced in this publication:

USAF JASU (For variations in technical data, refer to T.O. 35–1–7.)

ELECTRICAL STARTING UNITS:

A/M32A–86	AC: 115/200v, 3 phase, 90 kva, 0.8 pf, 4 wire
	DC: 28v, 1500 amp, 72 kw (with TR pack)
MC–1A	AC: 115/208v, 400 cycle, 3 phase, 37.5 kva, 0.8 pf, 108 amp, 4 wire
	DC: 28v, 500 amp, 14 kw
MD–3	AC: 115/208v, 400 cycle, 3 phase, 60 kva, 0.75 pf, 4 wire
	DC: 28v, 1500 amp, 45 kw, split bus
MD–3A	AC: 115/208v, 400 cycle, 3 phase, 60 kva, 0.75 pf, 4 wire
	DC: 28v, 1500 amp, 45 kw, split bus
MD–3M	AC: 115/208v, 400 cycle, 3 phase, 60 kva, 0.75 pf, 4 wire
	DC: 28v, 500 amp, 15 kw

NE, 09 FEB 2012 to 05 APR 2012

Legend 10. *Airport/facility directory.*

MD–4	AC: 120/208v, 400 cycle, 3 phase, 62.5 kva, 0.8 pf, 175 amp, "WYE" neutral ground, 4 wire, 120v, 400 cycle, 3 phase, 62.5 kva, 0.8 pf, 303 amp, "DELTA" 3 wire, 120v, 400 cycle, 1 phase, 62.5 kva, 0.8 pf, 520 amp, 2 wire

AIR STARTING UNITS

AM32–95	150 +/– 5 lb/min (2055 +/– 68 cfm) at 51 +/– 2 psia
AM32A–95	150 +/– 5 lb/min @ 49 +/– 2 psia (35 +/– 2 psig)
LASS	150 +/– 5 lb/min @ 49 +/– 2 psia
MA–1A	82 lb/min (1123 cfm) at 130° air inlet temp, 45 psia (min) air outlet press
MC–1	15 cfm, 3500 psia
MC–1A	15 cfm, 3500 psia
MC–2A	15 cfm, 200 psia
MC–11	8,000 cu in cap, 4000 psig, 15 cfm

COMBINED AIR AND ELECTRICAL STARTING UNITS:

AGPU	AC: 115/200v, 400 cycle, 3 phase, 30 kw gen
	DC: 28v, 700 amp
	AIR: 60 lb/min @ 40 psig @ sea level
AM32A–60*	AIR: 120 +/– 4 lb/min (1644 +/– 55 cfm) at 49 +/– 2 psia
	AC: 120/208v, 400 cycle, 3 phase, 75 kva, 0.75 pf, 4 wire, 120v, 1 phase, 25 kva
	DC: 28v, 500 amp, 15 kw
AM32A–60A	AIR: 150 +/– 5 lb/min (2055 +/– 68 cfm at 51 +/– psia
	AC: 120/208v, 400 cycle, 3 phase, 75 kva, 0.75 pf, 4 wire
	DC: 28v, 200 amp, 5.6 kw
AM32A–60B*	AIR: 130 lb/min, 50 psia
	AC: 120/208v, 400 cycle, 3 phase, 75 kva, 0.75 pf, 4 wire
	DC: 28v, 200 amp, 5.6 kw

*NOTE: During combined air and electrical loads, the pneumatic circuitry takes preference and will limit the amount of electrical power available.

USN JASU

ELECTRICAL STARTING UNITS:

NC–8A/A1	DC: 500 amp constant, 750 amp intermittent, 28v;
	AC: 60 kva @ .8 pf, 115/200v, 3 phase, 400 Hz.
NC–10A/A1/B/C	DC: 750 amp constant, 1000 amp intermittent, 28v;
	AC: 90 kva, 115/200v, 3 phase, 400 Hz.

AIR STARTING UNITS:

GTC–85/GTE–85	120 lbs/min @ 45 psi.
MSU–200NAV/A/U47A–5	204 lbs/min @ 56 psia.
WELLS AIR START SYSTEM	180 lbs/min @ 75 psi or 120 lbs/min @ 45 psi. Simultaneous multiple start capability.

COMBINED AIR AND ELECTRICAL STARTING UNITS:

NCPP–105/RCPT	180 lbs/min @ 75 psi or 120 lbs/min @ 45 psi. 700 amp, 28v DC. 120/208v, 400 Hz AC, 30 kva.

JASU (ARMY)

59B2–1B	28v, 7.5 kw, 280 amp.

OTHER JASU

ELECTRICAL STARTING UNITS (DND):

CE12	AC 115/200v, 140 kva, 400 Hz, 3 phase
CE13	AC 115/200v, 60 kva, 400 Hz, 3 phase
CE14	AC/DC 115/200v, 140 kva, 400 Hz, 3 phase, 28vDC, 1500 amp
CE15	DC 22–35v, 500 amp continuous 1100 amp intermittent
CE16	DC 22–35v, 500 amp continuous 1100 amp intermittent soft start

AIR STARTING UNITS (DND):

CA2	ASA 45.5 psig, 116.4 lb/min

COMBINED AIR AND ELECTRICAL STARTING UNITS (DND)

CEA1	AC 120/208v, 60 kva, 400 Hz, 3 phase DC 28v, 75 amp
	AIR 112.5 lb/min, 47 psig

ELECTRICAL STARTING UNITS (OTHER)

C–26	28v 45kw 115–200v 15kw 380–800 Hz 1 phase 2 wire
C–26–B, C–26–C	28v 45kw: Split Bus: 115–200v 15kw 380–800 Hz 1 phase 2 wire
E3	DC 28v/10kw

AIR STARTING UNITS (OTHER)

A4	40 psi/2 lb/sec (LPAS Mk12, Mk12L, Mk12A, Mk1, Mk2B)
MA–1	150 Air HP, 115 lb/min 50 psia
MA–2	250 Air HP, 150 lb/min 75 psia

CARTRIDGE:

MXU–4A	USAF

Legend 11. *Airport/facility directory.*

FUEL—MILITARY

Fuel available through US Military Base supply, DESC Into–Plane Contracts and/or reciprocal agreement is listed first and is followed by (Mil). At commercial airports where Into–Plane contracts are in place, the name of the refueling agent is shown. Military fuel should be used first if it is available. When military fuel cannot be obtained but Into–Plane contract fuel is available, Government aircraft must refuel with the contract fuel and applicable refueling agent to avoid any breach in contract terms and conditions. Fuel not available through the above is shown preceded by NC (no contract). When fuel is obtained from NC sources, local purchase procedures must be followed. The US Military Aircraft Identaplates DD Form 1896 (Jet Fuel), DD Form 1897 (Avgas) and AF Form 1245 (Avgas) are used at military installations only. The US Government Aviation Into–Plane Reimbursement (AIR) Card (currently issued by AVCARD) is the instrument to be used to obtain fuel under a DESC Into–Plane Contract and for NC purchases if the refueling agent at the commercial airport accepts the AVCARD. A current list of contract fuel locations is available online at www.desc.dla.mil/Static/ProductsAndServices.asp; click on the Commercial Airports button.

See legend item 14 for fuel code and description.

 SUPPORTING FLUIDS AND SYSTEMS—MILITARY

<u>CODE</u>

ADI	Anti–Detonation Injection Fluid—Reciprocating Engine Aircraft.
W	Water Thrust Augmentation—Jet Aircraft.
WAI	Water–Alcohol Injection Type, Thrust Augmentation—Jet Aircraft.
SP	Single Point Refueling.
PRESAIR	Air Compressors rated 3,000 PSI or more.
De–Ice	Anti–icing/De–icing/Defrosting Fluid (MIL–A–8243).

OXYGEN:

LPOX	Low pressure oxygen servicing.
HPOX	High pressure oxygen servicing.
LHOX	Low and high pressure oxygen servicing.
LOX	Liquid oxygen servicing.
OXRB	Oxygen replacement bottles. (Maintained primarily at Naval stations for use in acft where oxygen can be replenished only by replacement of cylinders.)
OX	Indicates oxygen servicing when type of servicing is unknown.

NOTE: Combinations of above items is used to indicate complete oxygen servicing available;

LHOXRB	Low and high pressure oxygen servicing and replacement bottles;
LPOXRB	Low pressure oxygen replacement bottles only, etc.

NOTE: Aircraft will be serviced with oxygen procured under military specifications only. Aircraft will not be serviced with medical oxygen.

NITROGEN:

LPNIT — Low pressure nitrogen servicing.
HPNIT — High pressure nitrogen servicing.
LHNIT — Low and high pressure nitrogen servicing.

 OIL—MILITARY

US AVIATION OILS (MIL SPECS):

CODE	GRADE, TYPE
O–113	1065, Reciprocating Engine Oil (MIL–L–6082)
O–117	1100, Reciprocating Engine Oil (MIL–L–6082)
O–117+	1100, O–117 plus cyclohexanone (MIL–L–6082)
O–123	1065, (Dispersant), Reciprocating Engine Oil (MIL–L–22851 Type III)
O–128	1100, (Dispersant), Reciprocating Engine Oil (MIL–L–22851 Type II)
O–132	1005, Jet Engine Oil (MIL–L–6081)
O–133	1010, Jet Engine Oil (MIL–L–6081)
O–147	None, MIL–L–6085A Lubricating Oil, Instrument, Synthetic
O–148	None, MIL–L–7808 (Synthetic Base) Turbine Engine Oil
O–149	None, Aircraft Turbine Engine Synthetic, 7.5c St
O–155	None, MIL–L–6086C, Aircraft, Medium Grade
O–156	None, MIL–L–23699 (Synthetic Base), Turboprop and Turboshaft Engines
JOAP/SOAP	Joint Oil Analysis Program. JOAP support is furnished during normal duty hours, other times on request. (JOAP and SOAP programs provide essentially the same service, JOAP is now the standard joint service supported program.)

 TRANSIENT ALERT (TRAN ALERT)—MILITARY

Tran Alert service is considered to include all services required for normal aircraft turn–around, e.g., servicing (fuel, oil, oxygen, etc.), debriefing to determine requirements for maintenance, minor maintenance, inspection and parking assistance of transient aircraft. Drag chute repack, specialized maintenance, or extensive repairs will be provided within the capabilities and priorities of the base. Delays can be anticipated after normal duty hours/holidays/weekends regardless of the hours of transient maintenance operation. Pilots should not expect aircraft to be serviced for TURN–AROUNDS during time periods when servicing or maintenance manpower is not available. In the case of airports not operated exclusively by US military, the servicing indicated by the remarks will not always be available for US military

NE, 09 FEB 2012 to 05 APR 2012

Legend 12. *Airport/facility directory.*

aircraft. When transient alert services are not shown, facilities are unknown. NO PRIORITY BASIS—means that transient alert services will be provided only after all the requirements for mission/tactical assigned aircraft have been accomplished.

 AIRPORT REMARKS

The Attendance Schedule is the months, days and hours the airport is actually attended. Airport attendance does not mean watchman duties or telephone accessibility, but rather an attendant or operator on duty to provide at least minimum services (e.g., repairs, fuel, transportation).

Airport Remarks have been grouped in order of applicability. Airport remarks are limited to those items of information that are determined essential for operational use, i.e., conditions of a permanent or indefinite nature and conditions that will remain in effect for more than 30 days concerning aeronautical facilities, services, maintenance available, procedures or hazards, knowledge of which is essential for safe and efficient operation of aircraft. Information concerning permanent closing of a runway or taxiway will not be shown. A note ''See Special Notices'' shall be applied within this remarks section when a special notice applicable to the entry is contained in the Special Notices section of this publication.

Parachute Jumping indicates parachute jumping areas associated with the airport. See Parachute Jumping Area section of this publication for additional Information.

Landing Fee indicates landing charges for private or non-revenue producing aircraft. In addition, fees may be charged for planes that remain over a couple of hours and buy no services, or at major airline terminals for all aircraft.

Note: Unless otherwise stated, remarks including runway ends refer to the runway's approach end.

 MILITARY REMARKS

Military Remarks published at a joint Civil/Military facility are remarks that are applicable to the Military. At Military Facilities all remarks will be published under the heading Military Remarks. Remarks contained in this section may not be applicable to civil users. The first group of remarks is applicable to the primary operator of the airport. Remarks applicable to a tenant on the airport are shown preceded by the tenant organization, i.e., (A) (AF) (N) (ANG), etc. Military airports operate 24 hours unless otherwise specified. Airport operating hours are listed first (airport operating hours will only be listed if they are different than the airport attended hours or if the attended hours are unavailable) followed by pertinent remarks in order of applicability. Remarks will include information on restrictions, hazards, traffic pattern, noise abatement, customs/agriculture/immigration, and miscellaneous information applicable to the Military.

Type of restrictions:

CLOSED: When designated closed, the airport is restricted from use by all aircraft unless stated otherwise. Any closure applying to specific type of aircraft or operation will be so stated. USN/USMC/USAF airports are considered closed during non–operating hours. Closed airports may be utilized during an emergency provided there is a safe landing area.

OFFICIAL BUSINESS ONLY: The airfield is closed to all transient military aircraft for obtaining routine services such as fueling, passenger drop off or pickup, practice approaches, parking, etc. The airfield may be used by aircrews and aircraft if official government business (including civilian) must be conducted on or near the airfield and prior permission is received from the airfield manager.

AF OFFICIAL BUSINESS ONLY OR NAVY OFFICIAL BUSINESS ONLY: Indicates that the restriction applies only to service indicated.

PRIOR PERMISSION REQUIRED (PPR): Airport is closed to transient aircraft unless approval for operation is obtained from the appropriate commander through Chief, Airfield Management or Airfield Operations Officer. Official Business or PPR does not preclude the use of US Military airports as an alternate for IFR flights. If a non–US military airport is used as a weather alternate and requires a PPR, the PPR must be requested and confirmed before the flight departs. The purpose of PPR is to control volume and flow of traffic rather than to prohibit it. Prior permission is required for all aircraft requiring transient alert service outside the published transient alert duty hours. All aircraft carrying hazardous materials must obtain prior permission as outlined in AFJI 11–204, AR 95–27, OPNAVINST 3710.7.

Note: OFFICIAL BUSINESS ONLY AND PPR restrictions are not applicable to Special Air Mission (SAM) or Special Air Resource (SPAR) aircraft providing person or persons on aboard are designated Code 6 or higher as explained in AFJMAN 11–213, AR 95–11, OPNAVINST 3722–8J. Official Business Only or PPR do not preclude the use of the airport as an alternate for IFR flights.

 WEATHER DATA SOURCES

Weather data sources will be listed alphabetically followed by their assigned frequencies and/or telephone number and hours of operation.

ASOS—Automated Surface Observing System. Reports the same as an AWOS–3 plus precipitation identification and intensity, and freezing rain occurrence;

AWOS—Automated Weather Observing System

  AWOS–A—reports altimeter setting (all other information is advisory only).

  AWOS–AV—reports altimeter and visibility.

  AWOS–1—reports altimeter setting, wind data and usually temperature, dew point and density altitude.

  AWOS–2—reports the same as AWOS–1 plus visibility.

  AWOS–3—reports the same as AWOS–1 plus visibility and cloud/ceiling data.

  AWOS–3P reports the same as the AWOS–3 system, plus a precipitation identification sensor.

  AWOS–3PT reports the same as the AWOS–3 system, plus precipitation identification sensor and a thunderstorm/lightning reporting capability.

  AWOS–3T reports the same as AWOS–3 system and includes a thunderstorm/lightning reporting capability.

NE, 09 FEB 2012 to 05 APR 2012

Legend 13. *Airport/facility directory.*

   See AIM, Basic Flight Information and ATC Procedures for detailed description of Weather Data Sources.

   AWOS–4—reports same as AWOS–3 system, plus precipitation occurence, type and accumulation, freezing rain, thunderstorm, and runway surface sensors.

HIWAS—See RADIO AIDS TO NAVIGATION

LAWRS—Limited Aviation Weather Reporting Station where observers report cloud height, weather, obstructions to vision, temperature and dewpoint (in most cases), surface wind, altimeter and pertinent remarks.

LLWAS—indicates a Low Level Wind Shear Alert System consisting of a center field and several field perimeter anemometers.

SAWRS—identifies airports that have a Supplemental Aviation Weather Reporting Station available to pilots for current weather information.

SWSL—Supplemental Weather Service Location providing current local weather information via radio and telephone.

TDWR—indicates airports that have Terminal Doppler Weather Radar.

WSP—indicates airports that have Weather System Processor.

When the automated weather source is broadcast over an associated airport NAVAID frequency (see NAVAID line), it shall be indicated by a bold ASOS, AWOS, or HIWAS followed by the frequency, identifier and phone number, if available.

 COMMUNICATIONS

Airport terminal control facilities and radio communications associated with the airport shall be shown. When the call sign is not the same as the airport name the call sign will be shown. Frequencies shall normally be shown in descending order with the primary frequency listed first. Frequencies will be listed, together with sectorization indicated by outbound radials, and hours of operation. Communications will be listed in sequence as follows:

Single Frequency Approach (SFA), Common Traffic Advisory Frequency (CTAF), Automatic Terminal Information Service (ATIS) and Aeronautical Advisory Stations (UNICOM) or (AUNICOM) along with their frequency is shown, where available, on the line following the heading ''COMMUNICATIONS.'' When the CTAF and UNICOM frequencies are the same, the frequency will be shown as CTAF/UNICOM 122.8.

The FSS telephone nationwide is toll free 1–800–WX–BRIEF (1–800–992–7433). When the FSS is located on the field it will be indicated as ''on arpt''. Frequencies available at the FSS will follow in descending order. Remote Communications Outlet (RCO) providing service to the airport followed by the frequency and FSS RADIO name will be shown when available.

FSS's provide information on airport conditions, radio aids and other facilities, and process flight plans. Airport Advisory Service (AAS) is provided on the CTAF by FSS's for select non-tower airports or airports where the tower is not in operation.

(See AIM, Para 4–1–9 Traffic Advisory Practices at Airports Without Operating Control Towers or AC 90–42C.)

Aviation weather briefing service is provided by FSS specialists. Flight and weather briefing services are also available by calling the telephone numbers listed.

Remote Communications Outlet (RCO)—An unmanned air/ground communications facility that is remotely controlled and provides UHF or VHF communications capability to extend the service range of an FSS.

Civil Communications Frequencies-Civil communications frequencies used in the FSS air/ground system are operated on 122.0, 122.2, 123.6; emergency 121.5; plus receive-only on 122.1.

    a. 122.0 is assigned as the Enroute Flight Advisory Service frequency at selected FSS RADIO outlets.

    b. 122.2 is assigned as a common enroute frequency.

    c. 123.6 is assigned as the airport advisory frequency at select non-tower locations. At airports with a tower, FSS may provide airport advisories on the tower frequency when tower is closed.

    d. 122.1 is the primary receive-only frequency at VOR's.

    e. Some FSS's are assigned 50 kHz frequencies in the 122–126 MHz band (eg. 122.45). Pilots using the FSS A/G system should refer to this directory or appropriate charts to determine frequencies available at the FSS or remoted facility through which they wish to communicate.

Emergency frequency 121.5 and 243.0 are available at all Flight Service Stations, most Towers, Approach Control and RADAR facilities.

Frequencies published followed by the letter ''T'' or ''R'', indicate that the facility will only transmit or receive respectively on that frequency. All radio aids to navigation (NAVAID) frequencies are transmit only.

TERMINAL SERVICES

SFA—Single Frequency Approach.

CTAF—A program designed to get all vehicles and aircraft at airports without an operating control tower on a common frequency.

ATIS—A continuous broadcast of recorded non-control information in selected terminal areas.

D–ATIS—Digital ATIS provides ATIS information in text form outside the standard reception range of conventional ATIS via landline & data link communications and voice message within range of existing transmitters.

AUNICOM—Automated UNICOM is a computerized, command response system that provides automated weather, radio check capability and airport advisory information selected from an automated menu by microphone clicks.

UNICOM—A non-government air/ground radio communications facility which may provide airport information.

PTD—Pilot to Dispatcher.

APP CON—Approach Control. The symbol ⓡ indicates radar approach control.

TOWER—Control tower.

GCA—Ground Control Approach System.

GND CON—Ground Control.

GCO—Ground Communication Outlet—An unstaffed, remotely controlled, ground/ground communications facility. Pilots at

NE, 09 FEB 2012 to 05 APR 2012

Legend 14. *Airport/facility directory.*

uncontrolled airports may contact ATC and FSS via VHF to a telephone connection to obtain an instrument clearance or close a VFR or IFR flight plan. They may also get an updated weather briefing prior to takeoff. Pilots will use four ''key clicks'' on the VHF radio to contact the appropriate ATC facility or six ''key clicks'' to contact the FSS. The GCO system is intended to be used only on the ground.

DEP CON—Departure Control. The symbol ® indicates radar departure control.

CLNC DEL—Clearance Delivery.

PRE TAXI CLNC—Pre taxi clearance.

VFR ADVSY SVC—VFR Advisory Service. Service provided by Non-Radar Approach Control.

Advisory Service for VFR aircraft (upon a workload basis) ctc APP CON.

COMD POST—Command Post followed by the operator call sign in parenthesis.

PMSV—Pilot-to-Metro Service call sign, frequency and hours of operation, when full service is other than continuous. PMSV installations at which weather observation service is available shall be indicated, following the frequency and/or hours of operation as ''Wx obsn svc 1900–0000Z‡'' or ''other times'' may be used when no specific time is given. PMSV facilities manned by forecasters are considered ''Full Service''. PMSV facilities manned by weather observers are listed as ''Limited Service''.

OPS—Operations followed by the operator call sign in parenthesis.

CON

RANGE

FLT FLW—Flight Following

MEDIVAC

NOTE: Communication frequencies followed by the letter ''X'' indicate frequency available on request.

33 AIRSPACE

Information concerning Class B, C, and part-time D and E surface area airspace shall be published with effective times. Class D and E surface area airspace that is continuous as established by Rulemaking Docket will not be shown.

CLASS B—Radar Sequencing and Separation Service for all aircraft in CLASS B airspace.

CLASS C—Separation between IFR and VFR aircraft and sequencing of VFR arrivals to the primary airport.

TRSA—Radar Sequencing and Separation Service for participating VFR Aircraft within a Terminal Radar Service Area.

Class C, D, and E airspace described in this publication is that airspace usually consisting of a 5 NM radius core surface area that begins at the surface and extends upward to an altitude above the airport elevation (charted in MSL for Class C and Class D). Class E surface airspace normally extends from the surface up to but not including the overlying controlled airspace.

When part-time Class C or Class D airspace defaults to Class E, the core surface area becomes Class E. This will be formatted as:

AIRSPACE: CLASS C svc ''times'' ctc APP CON other times CLASS E:

or

AIRSPACE: CLASS D svc ''times'' other times CLASS E.

When a part-time Class C, Class D or Class E surface area defaults to Class G, the core surface area becomes Class G up to, but not including, the overlying controlled airspace. Normally, the overlying controlled airspace is Class E airspace beginning at either 700′ or 1200′ AGL and may be determined by consulting the relevant VFR Sectional or Terminal Area Charts. This will be formatted as:

AIRSPACE: CLASS C svc ''times'' ctc APP CON other times CLASS G, with CLASS E 700′ (or 1200′) AGL & abv:

or

AIRSPACE: CLASS D svc ''times'' other times CLASS G with CLASS E 700′ (or 1200′) AGL & abv:

or

AIRSPACE: CLASS E svc ''times'' other times CLASS G with CLASS E 700′ (or 1200′) AGL & abv.

NOTE: AIRSPACE SVC "TIMES" INCLUDE ALL ASSOCIATED ARRIVAL EXTENSIONS. Surface area arrival extensions for instrument approach procedures become part of the primary core surface area. These extensions may be either Class D or Class E airspace and are effective concurrent with the times of the primary core surface area. For example, when a part-time Class C, Class D or Class E surface area defaults to Class G, the associated arrival extensions will default to Class G at the same time. When a part-time Class C or Class D surface area defaults to Class E, the arrival extensions will remain in effect as Class E airspace.

NOTE: CLASS E AIRSPACE EXTENDING UPWARD FROM 700 FEET OR MORE ABOVE THE SURFACE, DESIGNATED IN CONJUNCTION WITH AN AIRPORT WITH AN APPROVED INSTRUMENT PROCEDURE.

Class E 700′ AGL (shown as magenta vignette on sectional charts) and 1200′ AGL (blue vignette) areas are designated when necessary to provide controlled airspace for transitioning to/from the terminal and enroute environments. Unless otherwise specified, these 700′/1200′ AGL Class E airspace areas remain in effect continuously, regardless of airport operating hours or surface area status. These transition areas should not be confused with surface areas or arrival extensions.

(See Chapter 3, AIRSPACE, in the Aeronautical Information Manual for further details)

NE, 09 FEB 2012 to 05 APR 2012

Legend 15. *Airport/facility directory.*

34 RADIO AIDS TO NAVIGATION

The Airport/Facility Directory lists, by facility name, all Radio Aids to Navigation that appear on FAA, AeroNav Products Visual or IFR Aeronautical Charts and those upon which the FAA has approved an Instrument Approach Procedure, with exception of selected TACANs. Military TACAN information will be published for Military facilities contained in this publication. All VOR, VORTAC, TACAN, ILS and MLS equipment in the National Airspace System has an automatic monitoring and shutdown feature in the event of malfunction. Unmonitored, as used in this publication, for any navigational aid, means that monitoring personnel cannot observe the malfunction or shutdown signal. The NAVAID NOTAM file identifier will be shown as "NOTAM FILE IAD" and will be listed on the Radio Aids to Navigation line. When two or more NAVAIDS are listed and the NOTAM file identifier is different from that shown on the Radio Aids to Navigation line, it will be shown with the NAVAID listing. NOTAM file identifiers for ILSs and its components (e.g., NDB (LOM) are the same as the associated airports and are not repeated. Automated Surface Observing System (ASOS), Automated Weather Observing System (AWOS), and Hazardous Inflight Weather Advisory Service (HIWAS) will be shown when this service is broadcast over selected NAVAIDs.

NAVAID information is tabulated as indicated in the following sample:

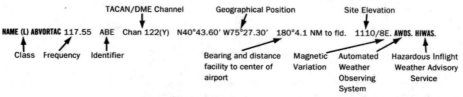

VOR unusable 020°–060° byd 26 NM blo 3,500'

Restriction within the normal altitude/range of the navigational aid (See primary alphabetical listing for restrictions on VORTAC and VOR/DME).

Note: Those DME channel numbers with a (Y) suffix require TACAN to be placed in the "Y" mode to receive distance information.

HIWAS—Hazardous Inflight Weather Advisory Service is a continuous broadcast of inflight weather advisories including summarized SIGMETs, convective SIGMETs, AIRMETs and urgent PIREPs. HIWAS is presently broadcast over selected VOR's throughout the U.S.

ASR/PAR—Indicates that Surveillance (ASR) or Precision (PAR) radar instrument approach minimums are published in the U.S. Terminal Procedures. Only part-time hours of operation will be shown.

RADIO CLASS DESIGNATIONS

VOR/DME/TACAN Standard Service Volume (SSV) Classifications

SSV Class	Altitudes	Distance (NM)
(T) Terminal	1000' to 12,000'	25
(L) Low Altitude	1000' to 18,000'	40
(H) High Altitude	1000' to 14,500'	40
	14,500' to 18,000'	100
	18,000' to 45,000'	130
	45,000' to 60,000'	100

NOTE: Additionally, (H) facilities provide (L) and (T) service volume and (L) facilities provide (T) service. Altitudes are with respect to the station's site elevation. Coverage is not available in a cone of airspace directly above the facility.

CONTINUED ON NEXT PAGE

NE, 09 FEB 2012 to 05 APR 2012

Legend 16. *Airport/facility directory.*

CONTINUED FROM PRECEDING PAGE

The term VOR is, operationally, a general term covering the VHF omnidirectional bearing type of facility without regard to the fact that the power, the frequency protected service volume, the equipment configuration, and operational requirements may vary between facilities at different locations.

AB	Automatic Weather Broadcast.
DF	Direction Finding Service.
DME	UHF standard (TACAN compatible) distance measuring equipment.
DME(Y)	UHF standard (TACAN compatible) distance measuring equipment that require TACAN to be placed in the ''Y'' mode to receive DME.
GS	Glide slope.
H	Non-directional radio beacon (homing), power 50 watts to less than 2,000 watts (50 NM at all altitudes).
HH	Non-directional radio beacon (homing), power 2,000 watts or more (75 NM at all altitudes).
H-SAB	Non-directional radio beacons providing automatic transcribed weather service.
ILS	Instrument Landing System (voice, where available, on localizer channel).
IM	Inner marker.
ISMLS	Interim Standard Microwave Landing System.
LDA	Localizer Directional Aid.
LMM	Compass locator station when installed at middle marker site (15 NM at all altitudes).
LOM	Compass locator station when installed at outer marker site (15 NM at all altitudes).
MH	Non-directional radio beacon (homing) power less than 50 watts (25 NM at all altitudes).
MLS	Microwave Landing System.
MM	Middle marker.
OM	Outer marker.
S	Simultaneous range homing signal and/or voice.
SABH	Non-directional radio beacon not authorized for IFR or ATC. Provides automatic weather broadcasts.
SDF	Simplified Direction Facility.
TACAN	UHF navigational facility-omnidirectional course and distance information.
VOR	VHF navigational facility-omnidirectional course only.
VOR/DME	Collocated VOR navigational facility and UHF standard distance measuring equipment.
VORTAC	Collocated VOR and TACAN navigational facilities.
W	Without voice on radio facility frequency.
Z	VHF station location marker at a LF radio facility.

NE, 09 FEB 2012 to 05 APR 2012

Legend 17. *Airport/facility directory.*

DIRECTORY LEGEND
ILS FACILITY PEFORMANCE CLASSIFICATION CODES

Codes define the ability of an ILS to support autoland operations. The two portions of the code represent Official Category and farthest point along a Category I, II, or III approach that the Localizer meets Category III structure tolerances.

Official Category: I, II, or III; the lowest minima on published or unpublished procedures supported by the ILS.

Farthest point of satisfactory Category III Localizer performance for Category I, II, or III approaches: A – 4 NM prior to runway threshold, B – 3500 ft prior to runway threshold, C – glide angle dependent but generally 750–1000 ft prior to threshold, T – runway threshold, D – 3000 ft after runway threshold, and E – 2000 ft prior to stop end of runway.

ILS information is tabulated as indicated in the following sample:

ILS/DME   108.5   I–ORL   Chan 22   Rwy 18.   Class IIE.   LOM HERNY NDB.

ILS Facility Performance
Classification Code

FREQUENCY PAIRING PLAN AND MLS CHANNELING

MLS CHANNEL	VHF FREQUENCY	TACAN CHANNEL	MLS CHANNEL	VHF FREQUENCY	TACAN CHANNEL	MLS CHANNEL	VHF FREQUENCY	TACAN CHANNEL
500	108.10	18X	568	109.45	31Y	636	114.15	88Y
502	108.30	20X	570	109.55	32Y	638	114.25	89Y
504	108.50	22X	572	109.65	33Y	640	114.35	90Y
506	108.70	24X	574	109.75	34Y	642	114.45	91Y
508	108.90	26X	576	109.85	35Y	644	114.55	92Y
510	109.10	28X	578	109.95	36Y	646	114.65	93Y
512	109.30	30X	580	110.05	37Y	648	114.75	94Y
514	109.50	32X	582	110.15	38Y	650	114.85	95Y
516	109.70	34X	584	110.25	39Y	652	114.95	96Y
518	109.90	36X	586	110.35	40Y	654	115.05	97Y
520	110.10	38X	588	110.45	41Y	656	115.15	98Y
522	110.30	40X	590	110.55	42Y	658	115.25	99Y
524	110.50	42X	592	110.65	43Y	660	115.35	100Y
526	110.70	44X	594	110.75	44Y	662	115.45	101Y
528	110.90	46X	596	110.85	45Y	664	115.55	102Y
530	111.10	48X	598	110.95	46Y	666	115.65	103Y
532	111.30	50X	600	111.05	47Y	668	115.75	104Y
534	111.50	52X	602	111.15	48Y	670	115.85	105Y
536	111.70	54X	604	111.25	49Y	672	115.95	106Y
538	111.90	56X	606	111.35	50Y	674	116.05	107Y
540	108.05	17Y	608	111.45	51Y	676	116.15	108Y
542	108.15	18Y	610	111.55	52Y	678	116.25	109Y
544	108.25	19Y	612	111.65	53Y	680	116.35	110Y
546	108.35	20Y	614	111.75	54Y	682	116.45	111Y
548	108.45	21Y	616	111.85	55Y	684	116.55	112Y
550	108.55	22Y	618	111.95	56Y	686	116.65	113Y
552	108.65	23Y	620	113.35	80Y	688	116.75	114Y
554	108.75	24Y	622	113.45	81Y	690	116.85	115Y
556	108.85	25Y	624	113.55	82Y	692	116.95	116Y
558	108.95	26Y	626	113.65	83Y	694	117.05	117Y
560	109.05	27Y	628	113.75	84Y	696	117.15	118Y
562	109.15	28Y	630	113.85	85Y	698	117.25	119Y
564	109.25	29Y	632	113.95	86Y			
566	109.35	30Y	634	114.05	87Y			

FREQUENCY PAIRING PLAN AND MLS CHANNELING

The following is a list of paired VOR/ILS VHF frequencies with TACAN channels and MLS channels.

TACAN CHANNEL	VHF FREQUENCY	MLS CHANNEL	TACAN CHANNEL	VHF FREQUENCY	MLS CHANNEL	TACAN CHANNEL	VHF FREQUENCY	MLS CHANNEL
2X	134.5	-	19Y	108.25	544	25X	108.80	-
2Y	134.55	-	20X	108.30	502	25Y	108.85	556
11X	135.4	-	20Y	108.35	546	26X	108.90	508
11Y	135.45	-	21X	108.40	-	26Y	108.95	558
12X	135.5	-	21Y	108.45	548	27X	109.00	-
12Y	135.55	-	22X	108.50	504	27Y	109.05	560
17X	108.00	-	22Y	108.55	550	28X	109.10	510
17Y	108.05	540	23X	108.60	-	28Y	109.15	562
18X	108.10	500	23Y	108.65	552	29X	109.20	-
18Y	108.15	542	24X	108.70	506	29Y	109.25	564
19X	108.20	-	24Y	108.75	554	30X	109.30	512

NE, 09 FEB 2012 to 05 APR 2012

Legend 18. *Airport/facility directory.*

TACAN CHANNEL	VHF FREQUENCY	MLS CHANNEL	TACAN CHANNEL	VHF FREQUENCY	MLS CHANNEL	TACAN CHANNEL	VHF FREQUENCY	MLS CHANNEL
30Y	109.35	566	63X	133.60	-	95Y	114.85	650
31X	109.40	-	63Y	133.65	-	96X	114.90	-
31Y	109.45	568	64X	133.70	-	96Y	114.95	652
32X	109.50	514	64Y	133.75	-	97X	115.00	-
32Y	109.55	570	65X	133.80	-	97Y	115.05	654
33X	109.60	-	65Y	133.85	-	98X	115.10	-
33Y	109.65	572	66X	133.90	-	98Y	115.15	656
34X	109.70	516	66Y	133.95	-	99X	115.20	-
34Y	109.75	574	67X	134.00	-	99Y	115.25	658
35X	109.80	-	67Y	134.05	-	100X	115.30	-
35Y	109.85	576	68X	134.10	-	100Y	115.35	660
36X	109.90	518	68Y	134.15	-	101X	115.40	-
36Y	109.95	578	69X	134.20	-	101Y	115.45	662
37X	110.00	-	69Y	134.25	-	102X	115.50	-
37Y	110.05	580	70X	112.30	-	102Y	115.55	664
38X	110.10	520	70Y	112.35	-	103X	115.60	-
38Y	110.15	582	71X	112.40	-	103Y	115.65	666
39X	110.20	-	71Y	112.45	-	104X	115.70	-
39Y	110.25	584	72X	112.50	-	104Y	115.75	668
40X	110.30	522	72Y	112.55	-	105X	115.80	-
40Y	110.35	586	73X	112.60	-	105Y	115.85	670
41X	110.40	-	73Y	112.65	-	106X	115.90	-
41Y	110.45	588	74X	112.70	-	106Y	115.95	672
42X	110.50	524	74Y	112.75	-	107X	116.00	-
42Y	110.55	590	75X	112.80	-	107Y	116.05	674
43X	110.60	-	75Y	112.85	-	108X	116.10	-
43Y	110.65	592	76X	112.90	-	108Y	116.15	676
44X	110.70	526	76Y	112.95	-	109X	116.20	-
44Y	110.75	594	77X	113.00	-	109Y	116.25	678
45X	110.80	-	77Y	113.05	-	110X	116.30	-
45Y	110.85	596	78X	113.10	-	110Y	116.35	680
46X	110.90	528	78Y	113.15	-	111X	116.40	-
46Y	110.95	598	79X	113.20	-	111Y	116.45	682
47X	111.00	-	79Y	113.25	-	112X	116.50	-
47Y	111.05	600	80X	113.30	-	112Y	116.55	684
48X	111.10	530	80Y	113.35	620	113X	116.60	-
48Y	111.15	602	81X	113.40	-	113Y	116.65	686
49X	111.20	-	81Y	113.45	622	114X	116.70	-
49Y	111.25	604	82X	113.50	-	114Y	116.75	688
50X	111.30	532	82Y	113.55	624	115X	116.80	-
50Y	111.35	606	83X	113.60	-	115Y	116.85	690
51X	111.40	-	83Y	113.65	626	116X	116.90	-
51Y	111.45	608	84X	113.70	-	116Y	116.95	692
52X	111.50	534	84Y	113.75	628	117X	117.00	-
52Y	111.55	610	85X	113.80	-	117Y	117.05	694
53X	111.60	-	85Y	113.85	630	118X	117.10	-
53Y	111.65	612	86X	113.90	-	118Y	117.15	696
54X	111.70	536	86Y	113.95	632	119X	117.20	-
54Y	111.75	614	87X	114.00	-	119Y	117.25	698
55X	111.80	-	87Y	114.05	634	120X	117.30	-
55Y	111.85	616	88X	114.10	-	120Y	117.35	-
56X	111.90	538	88Y	114.15	636	121X	117.40	-
56Y	111.95	618	89X	114.20	-	121Y	117.45	-
57X	112.00	-	89Y	114.25	638	122X	117.50	-
57Y	112.05	-	90X	114.30	-	122Y	117.55	-
58X	112.10	-	90Y	114.35	640	123X	117.60	-
58Y	112.15	-	91X	114.40	-	123Y	117.65	-
59X	112.20	-	91Y	114.45	642	124X	117.70	-
59Y	112.25	-	92X	114.50	-	124Y	117.75	-
60X	133.30	-	92Y	114.55	644	125X	117.80	-
60Y	133.35	-	93X	114.60	-	125Y	117.85	-
61X	133.40	-	93Y	114.65	646	126X	117.90	-
61Y	133.45	-	94X	114.70	-	126Y	117.95	-
62X	133.50	-	94Y	114.75	648			
62Y	133.55	-	95X	114.80	-			

35 COMM/NAV/WEATHER REMARKS:

These remarks consist of pertinent information affecting the current status of communications, NAVAIDs and weather.

NE, 09 FEB 2012 to 05 APR 2012

Legend 19. *Airport/facility directory.*

Appendix 2

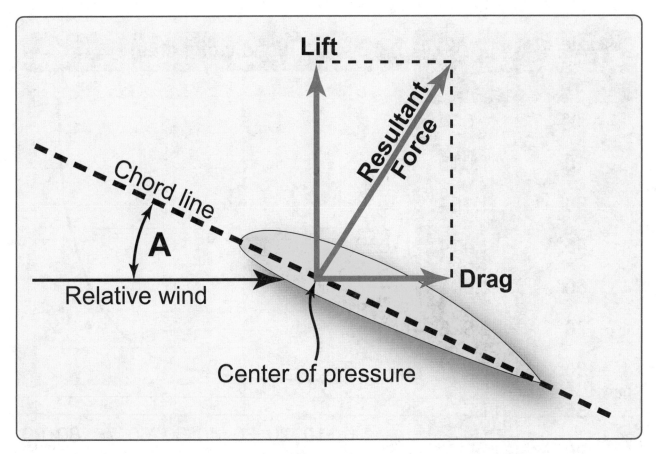

Figure 1. *Lift vector.*

Angle of bank ϕ	Load factor n
0°	1.0
10°	1.015
30°	1.154
45°	1.414
60°	2.000
70°	2.923
80°	5.747
85°	11.473
90°	∞

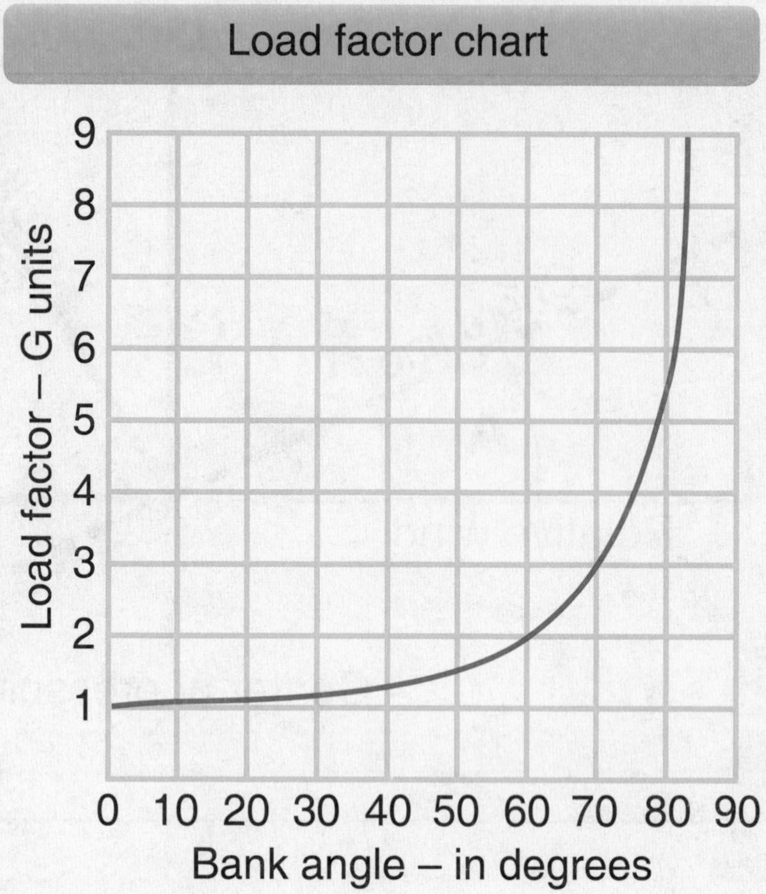

Figure 2. *Load factor chart.*

Figure 3. *Altimeter.*

5

Figure 4. *Airspeed indicator.*

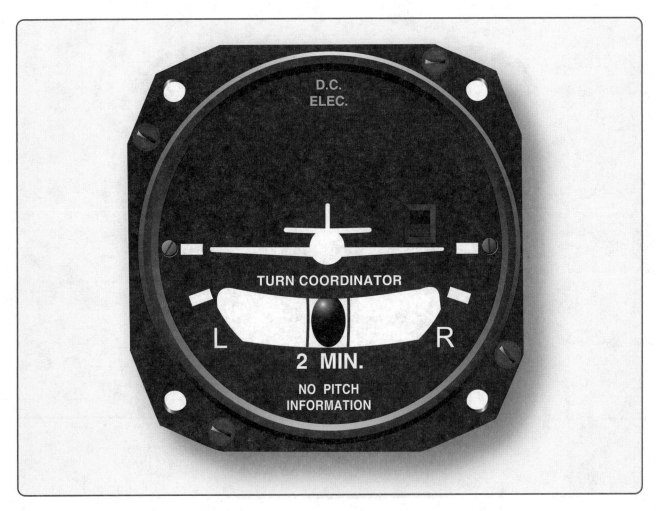

Figure 5. *Turn coordinator.*

Figure 6. *Heading indicator.*

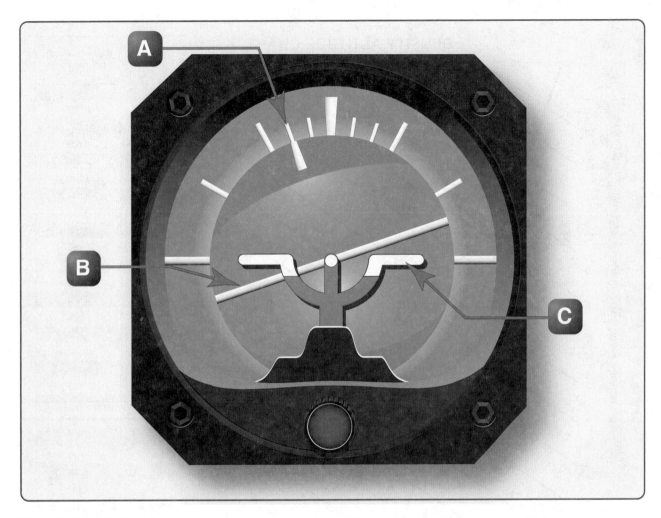

Figure 7. *Attitude indicator.*

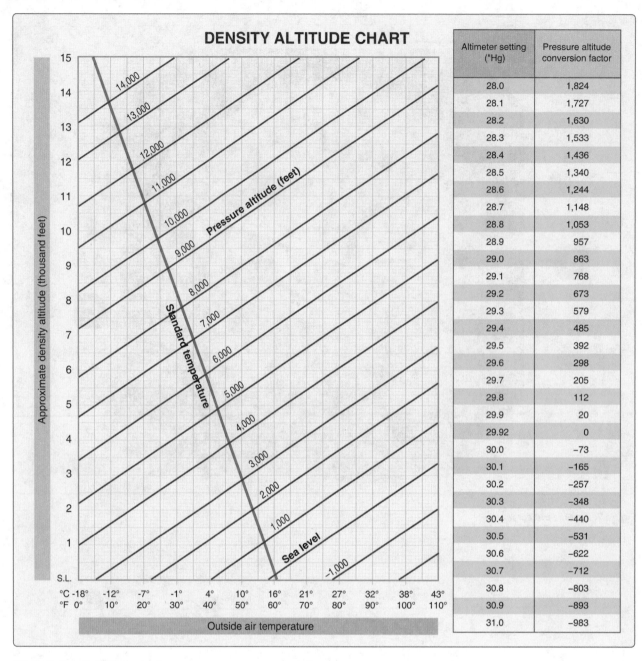

Figure 8. *Density altitude chart.*

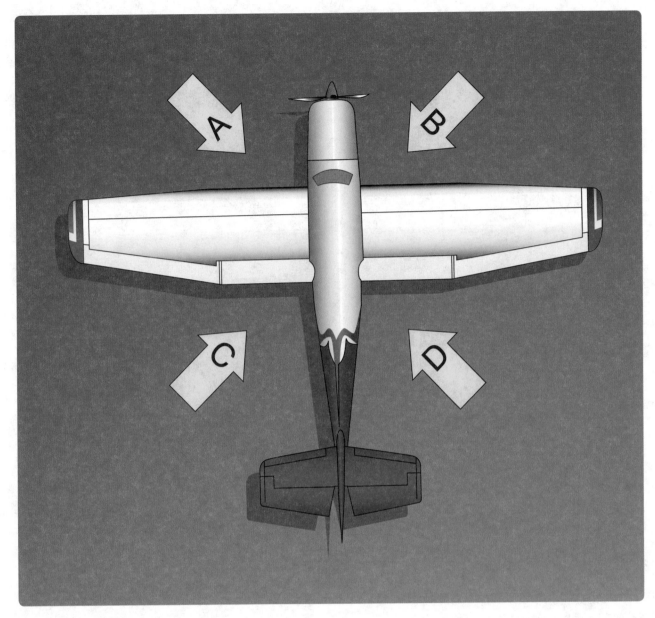

Figure 9. *Control position for taxi.*

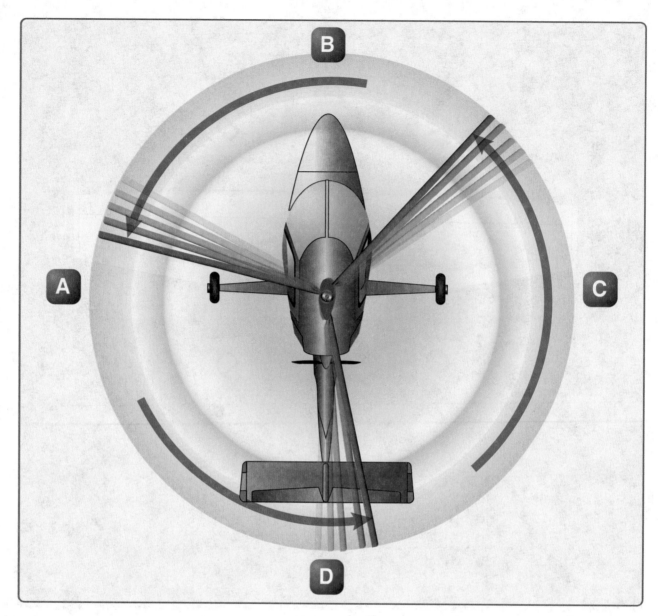

Figure 10. *Gyroplane rotor blade position.*

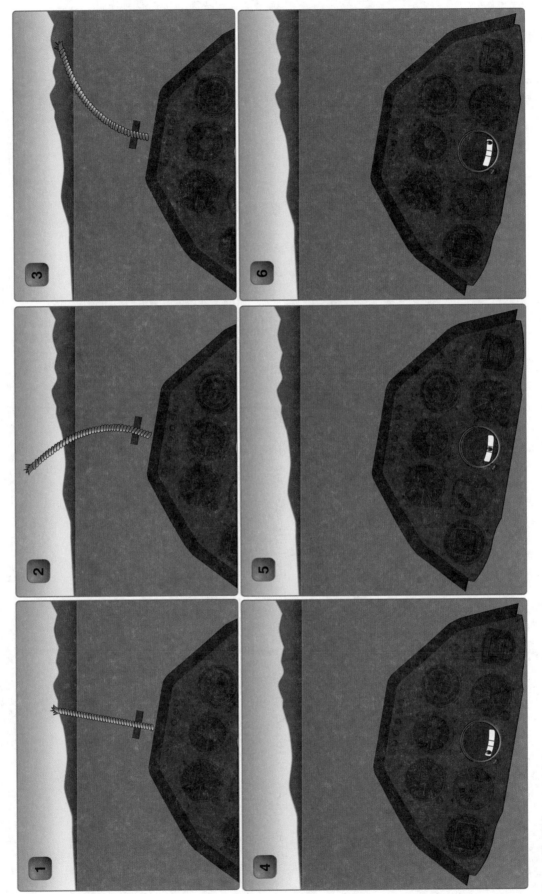

Figure 11. *Glider yaw string.*

METAR KINK 121845Z 11012G18KT 15SM SKC 25/17 A3000

METAR KBOI 121854Z 13004KT 30SM SCT150 17/6 A3015

METAR KLAX 121852Z 25004KT 6SM BR SCT007 SCT250 16/15 A2991

SPECI KMDW 121856Z 32005KT 1 1/2SM RA OVC007 17/16 A2980 RMK RAB35

SPECI KJFK 121853Z 18004KT 1/2SM FG R04/2200 OVC005 20/18 A3006

Figure 12. *Aviation routine weather reports (METAR).*

14

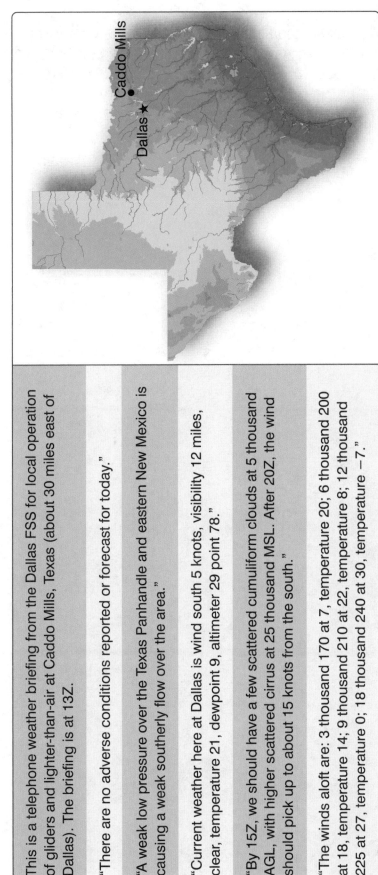

This is a telephone weather briefing from the Dallas FSS for local operation of gliders and lighter-than-air at Caddo Mills, Texas (about 30 miles east of Dallas). The briefing is at 13Z.

"There are no adverse conditions reported or forecast for today."

"A weak low pressure over the Texas Panhandle and eastern New Mexico is causing a weak southerly flow over the area."

"Current weather here at Dallas is wind south 5 knots, visibility 12 miles, clear, temperature 21, dewpoint 9, altimeter 29 point 78."

"By 15Z, we should have a few scattered cumuliform clouds at 5 thousand AGL, with higher scattered cirrus at 25 thousand MSL. After 20Z, the wind should pick up to about 15 knots from the south."

"The winds aloft are: 3 thousand 170 at 7, temperature 20; 6 thousand 200 at 18, temperature 14; 9 thousand 210 at 22, temperature 8; 12 thousand 225 at 27, temperature 0; 18 thousand 240 at 30, temperature −7."

Figure 13. *Telephone weather briefing.*

UA/OV KOKC-KTUL/TM 1800/FL 120/TP BE90//SK BKN018-TOP055/OVC072-TOP089/CLR ABV/TA M7/WV 08021/TB LGT 055-072/IC LGT-MOD RIME 072-089

Figure 14. *Pilot weather report.*

TAF	
KMEM	121720Z 121818 20012KT 5SM HZ BKN030 PROB40 2022 1SM TSRA OVC008CB FM2200 33015G20KT P6SM BKN015 OVC025 PROB40 2202 3SM SHRA FM0200 35012KT OVC008 PROB40 0205 2SM-RASN BECMG 0608 02008KT BKN012 BECMG 1012 00000KT 3SM BR SKC TEMPO 1214 1/2SM FG FM1600 VRB06KT P6SM SKC=
KOKC	051130Z 051212 14008KT 5SM BR BKN030 TEMPO 1316 1 1/2SM BR FM1600 18010KT P6SM SKC BECMG 2224 20013G20KT 4SM SHRA OVC020 PROB40 0006 2SM TSRA OVC008CB BECMG 0608 21015KT P6SM SCT040=

Figure 15. *Terminal aerodrome forecasts (TAF).*

BOSC FA 241845
SYNOPSIS AND VFR CLDS/WX
SYNOPSIS VALID UNTIL 251300
CLDS/WX VALID UNTIL 250700...OTLK VALID 250700-251300
ME NH VT MA RI CT NY LO NJ PA OH LE WV MD DC DE VA AND CSTL WTRS
.
SEE AIRMET SIERRA FOR IFR CONDS AND MTN OBSCN.
TS IMPLY SEV OR GTR TURB SEV ICE LLWS AND IFR CONDS.
NON MSL HGTS DENOTED BY AGL OR CIG.
.
SYNOPSIS...19Z CDFNT ALG A 16NE ACK-ENE LN...CONTG AS A QSTNRY
FNT ALG AN END-50SW MSS LN. BY 13Z...CDFNT ALG A 140ESE ACK-HTO
LN...CONTG AS A QSTNRY FNT ALG A HTO-SYR-YYZ LN. TROF ACRS CNTRL
PA INTO NRN VA. ...REYNOLDS...
.
OH LE
NRN HLF OH LE...SCT-BKN025 OVC045. CLDS LYRD 150. SCT SHRA. WDLY
      SCT TSRA. CB TOPS FL350. 23-01Z OVC020-030. VIS 3SM BR. OCNL-
      RA. OTLK...IFR CIG BR FG.
SWRN QTR OH...BKN050-060 TOPS 100. OTLK...MVFR BR.
SERN QTR OH...SCT-BKN040 BKN070 TOPS 120. WDLY SCT-TSRA. 00Z
      SCT-BKN030 OVC050. WDLY SCT-TSRA. CB TOPS FL350. OTLK...VFR
      SHRA.
.
CHIC FA 241945
SYNOPSIS AND VFR CLDS/WX
SYNOPSIS VALID UNTIL 251400
CLDS/WX VALID UNTIL 250800...OTLK VALID 250800-251400
ND SD NE KS MN IA MO WI LM LS MI LH IL IN KY
.
SEE AIRMET SIERRA FOR IFR CONDS AND MTN OBSCN.
TS IMPLY SEV OR GTR TURB SEV ICE LLWS AND IFR CONDS.
NON MSL HGTS DENOTED BY AGL OR CIG.
.
SYNOPSIS...LOW PRES AREA 20Z CNTRD OVR SERN WI FCST MOV NEWD INTO
LH BY 12Z AND WKN. LOW PRES FCST DEEPEN OVR ERN CO DURG PD AND
MOV NR WRN KS BORDER BY 14Z. DVLPG CDFNT WL MOV EWD INTO S CNTRL
NE-CNTRL KS BY 14Z. ...SMITH..
.
UPR MI LS
WRN PTNS...AGL SCT030 SCT 030 SCT-BKN050. TOPS 080. 02-05Z BECMG CIG
      OVC010 VIS 3-5SM BR. OTLK...IFR CIG BR.
ERN PTNS...CIG BKN020 OVC040. OCNL VIS 3-5SM -RA BR. TOPS FL200.
      23Z CIG OVC010 VIS 3-5SM -RA BR. OTLK...IFR CIG BR.
.
LWR MI LM LH
CNTRL/NRN PTNS...CIG OVC010 VIS 3-5SM -RA BR. TOPS FL200
      OTLK...IFR CIG BR.
.
SRN THIRD...CIG OVC015-025. SCT -SHRA. TOPS 150. 00-02Z BECMG CIG
      OVC010 VIS 3-5SM BR. TOPS 060. OTLK...IFR CIG BR.
.
IN
NRN HALF...CIG BKN035 BKN080. TOPS FL200. SCT -SHRA. 00Z CIG
      BKN-SCT040 BKN-SCT080. TOPS 120. 06Z AGL SCT-BKN030. TOPS 080.
      OCNL VIS 3-5SM BR. OTLK...MVFR CIG BR.
SRN HALF...AGL SCT050 SCT-BKN100. TOPS 120. 07Z AGL SCT 030
      SCT100. OTLK...VFR.

Figure 16. *Area forecast.*

FD WBC 151745
DATA BASED ON 151200Z
VALID 1600Z FOR USE 1800-0300Z. TEMPS NEG ABV 24000

FT	3000	6000	9000	12000	18000	24000	30000	34000	39000
ALS			2420	2635-08	2535-18	2444-30	245945	246755	246862
AMA		2714	2725+00	2625-04	2531-15	2542-27	265842	256352	256762
DEN			2321-04	2532-08	2434-19	2441-31	235347	236056	236262
HLC		1707-01	2113-03	2219-07	2330-17	2435-30	244145	244854	245561
MKC	0507	2006+03	2215-01	2322-06	2338-17	2348-29	236143	237252	238160
STL	2113	2325+07	2332+02	2339-04	2356-16	2373-27	239440	730649	731960

Figure 17. *Winds and temperatures aloft forecast.*

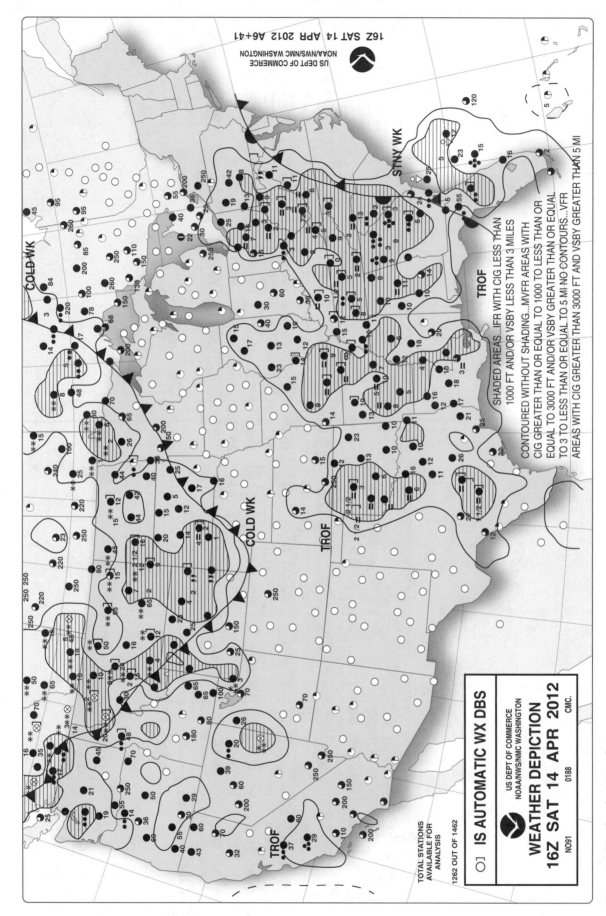

Figure 18. *Weather depiction chart.*

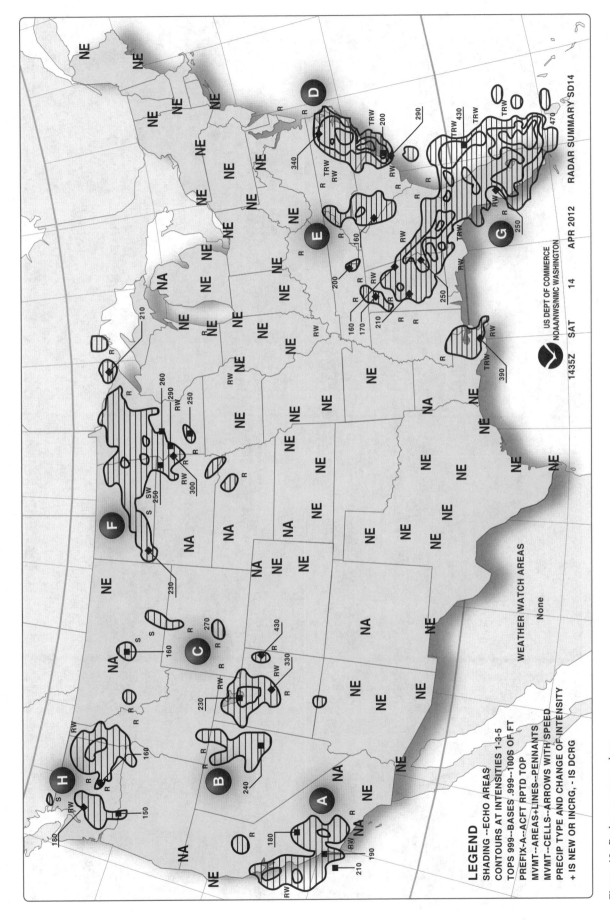

Figure 19. *Radar summary chart.*

21

Figure 20. *Low-level significant weather (SIGWX) prognostic charts.*

22

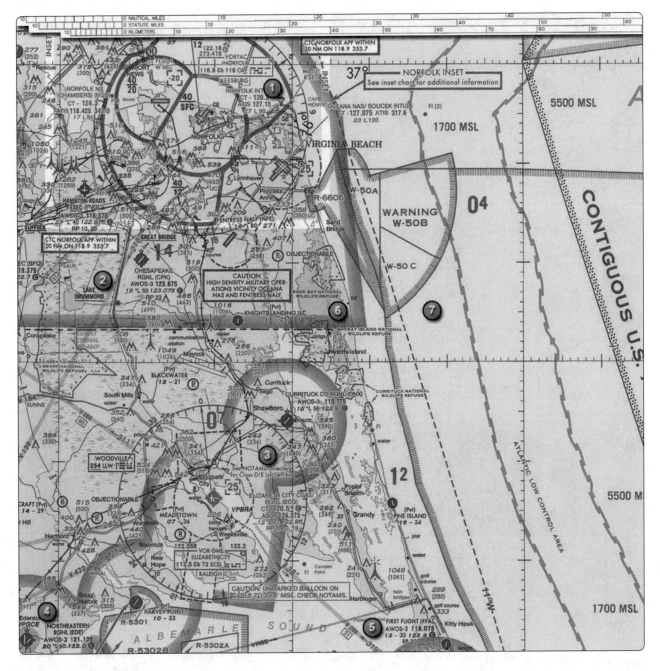

Figure 21. *Sectional chart excerpt.*

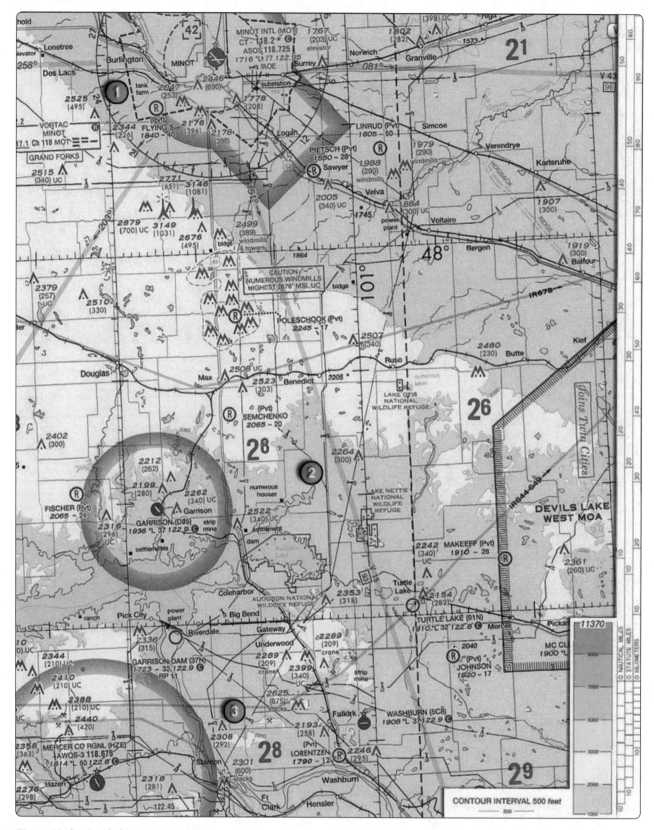

Figure 22. *Sectional chart excerpt.*

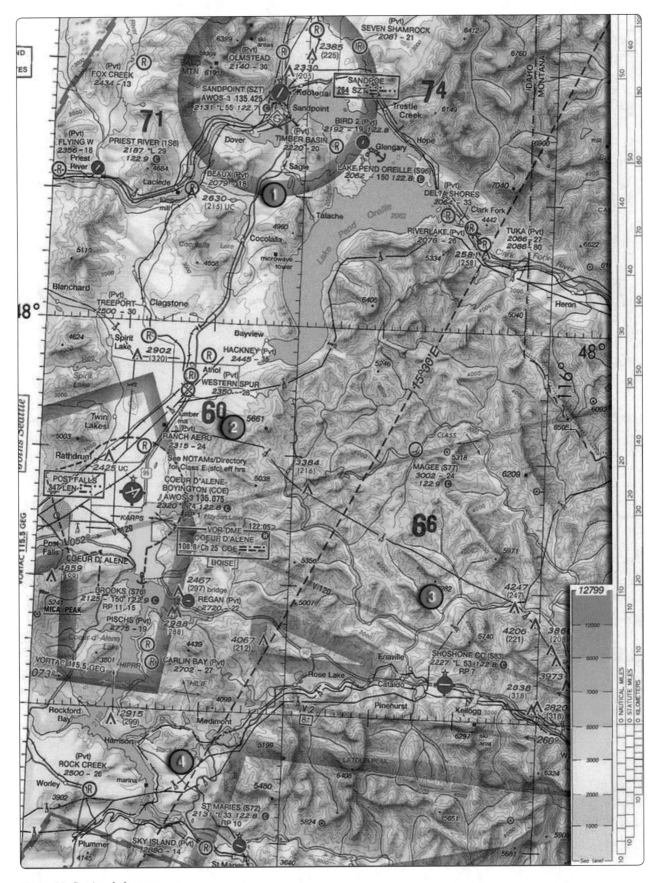

Figure 23. *Sectional chart excerpt.*

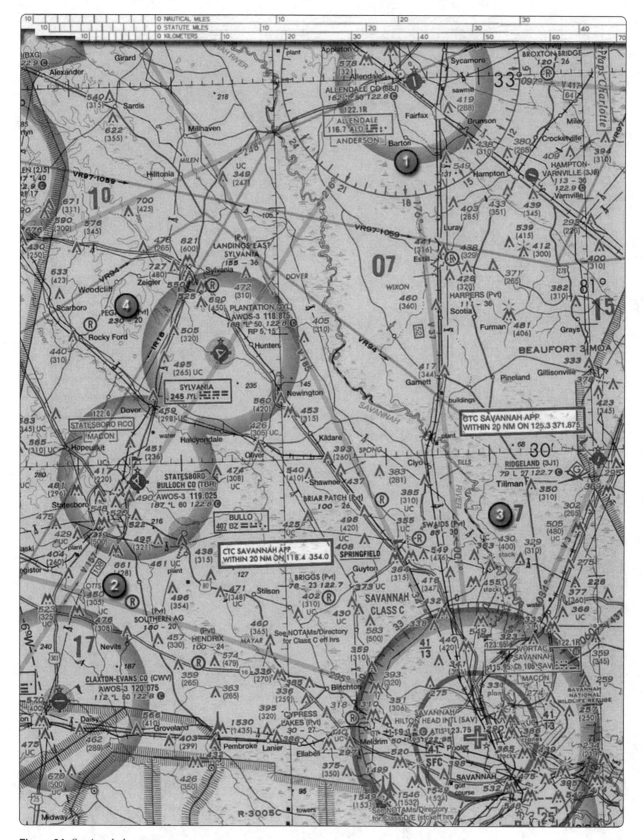

Figure 24. *Sectional chart excerpt.*

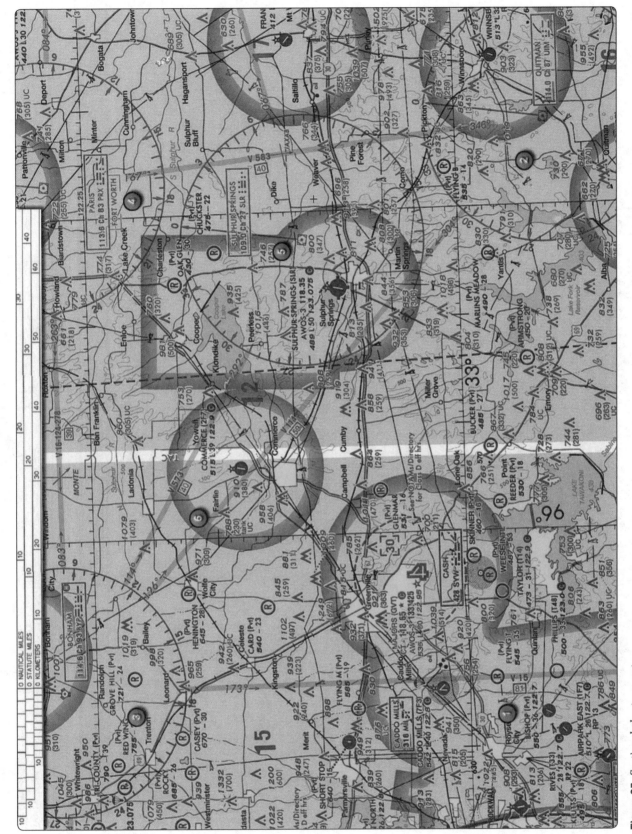

Figure 25. *Sectional chart excerpt.*

27

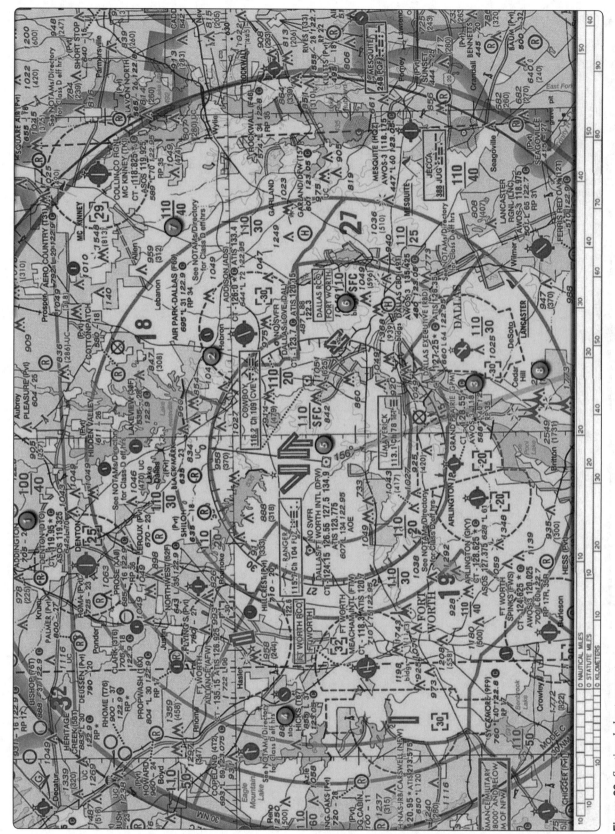

Figure 26. *Sectional chart excerpt.*

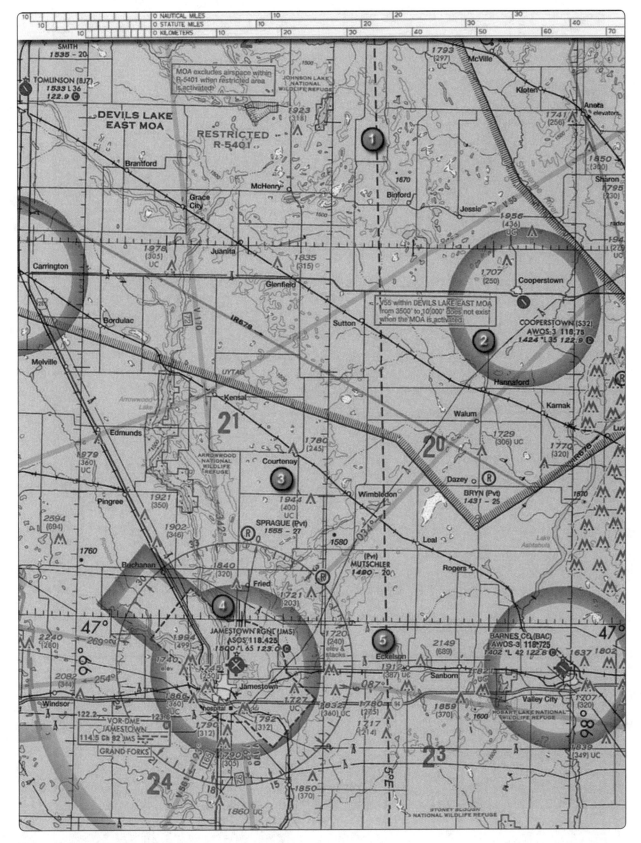

Figure 27. *Sectional chart excerpt.*

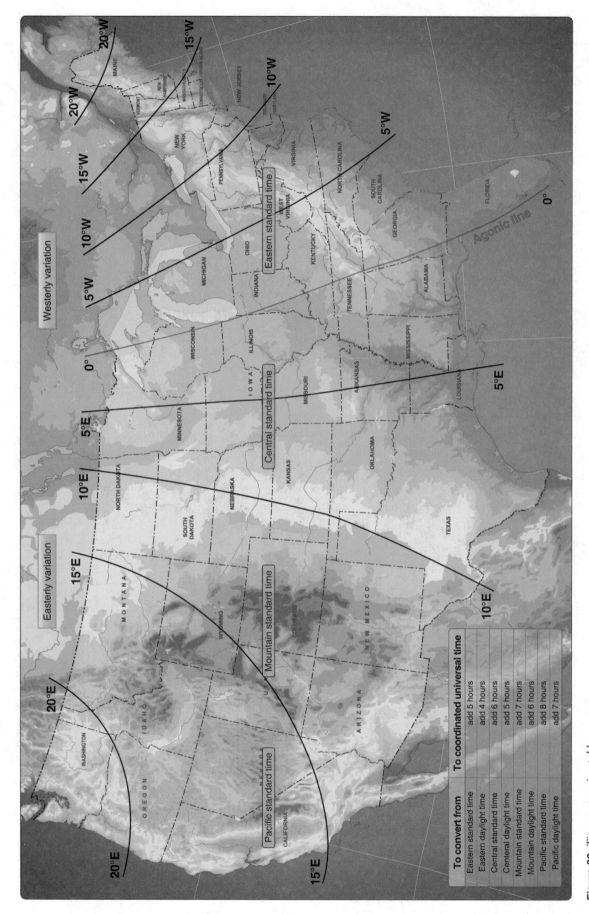

Figure 28. *Time conversion table.*

To convert from	To coordinated universal time
Eastern standard time	add 5 hours
Eastern daylight time	add 4 hours
Central standard time	add 6 hours
Centeral daylight time	add 5 hours
Mountain standard time	add 7 hours
Mountain daylight time	add 6 hours
Pacific standard time	add 8 hours
Pacific daylight time	add 7 hours

30

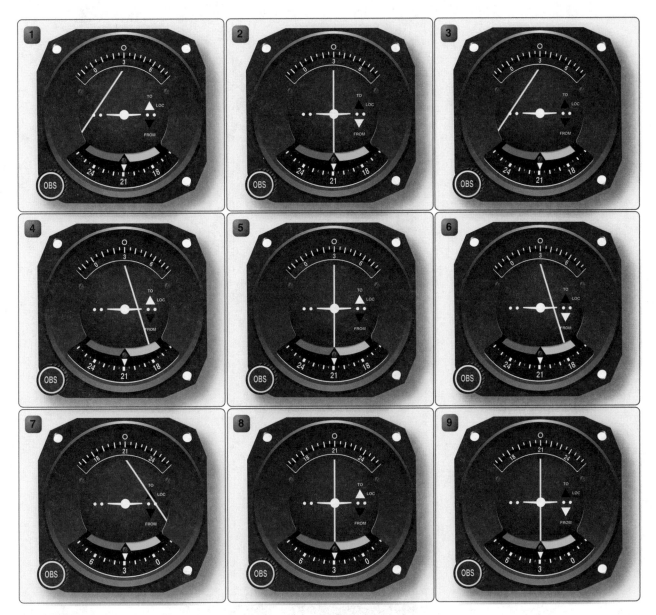

Figure 29. *VOR.*

Figure 30. *ADF (movable card).*

32

Figure 31. *ADF (fixed card).*

33

COEUR D'ALENE–PAPPY BOYINGTON FLD (COE) 9 NW UTC–8(–7DT)

N47°46.46' W116°49.18'

2320 B S4 **FUEL** 100, JET A OX 1, 2, 3, 4 Class IV, ARFF Index A NOTAM FILE COE

RWY 05–23: H7400X100 (ASPH-GRVD) S–57, D–95, 2S–121, 2D–165 HIRL 0.6% up NE

RWY 05: MALSR (NSTD). PAPI(P4R)—GA 3.0° TCH 56'.

RWY 23: REIL. PAPI(P4R)—GA 3.0° TCH 50'.

RWY 01–19: H5400X75 (ASPH) S–50, D–83, 2S–105, 2D–150

MIRL 0.3% up N

RWY 01: REIL. PAPI(P2L)—GA 3.0° TCH 39'. Rgt tfc.

RWY 19: PAPI(P2L)—GA 3.0° TCH 41'.

RUNWAY DECLARED DISTANCE INFORMATION

	TORA	TODA	ASDA	LDA
RWY 01:	TORA–5400	TODA–5400	ASDA–5400	LDA–5400
RWY 05:	TORA–7400	TODA–7400	ASDA–7400	LDA–7400
RWY 19:	TORA–5400	TODA–5400	ASDA–5400	LDA–5400
RWY 23:	TORA–7400	TODA–7400	ASDA–7400	LDA–7400

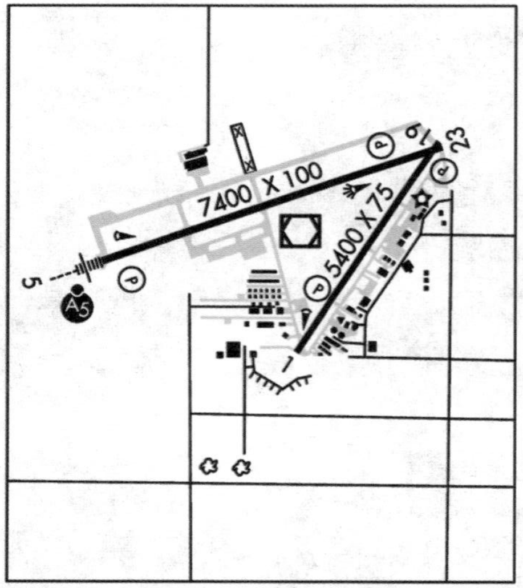

AIRPORT REMARKS: Attended Mon–Fri 1500–0100Z‡. For after hrs fuel-self svc avbl or call 208-772-6404, 208-661-4174, 208-661-7449, 208-699-5433. Self svc fuel avbl with credit card. 48 hr PPR for unscheduled ops with more than 30 passenger seats call arpt manager 208-446-1860. Migratory birds on and invof arpt Oct–Nov. Remote cntl airstrip is 2.3 miles west AER 05. Arpt conditions avbl on AWOS. Rwy 05 NSTD MALSR, thld bar extends 5' byd rwy edge lgts each side. ACTIVATE MIRL Rwy 01–19, HIRL Rwy 05–23, REIL Rwy 01 and Rwy 23, MALSR Rwy 05—CTAF. PAPI Rwy 01, Rwy 19, Rwy 05, and Rwy 23 opr continuously.

WEATHER DATA SOURCES: AWOS–3 135.075 (208) 772–8215.

HIWAS 108.8 COE.

COMMUNICATIONS: CTAF/UNICOM 122.8

RCO 122.05 (BOISE RADIO)

Ⓡ **SPOKANE APP/DEP CON** 132.1

AIRSPACE: CLASS E svc continuous.

RADIO AIDS TO NAVIGATION: NOTAM FILE COE.

(T) VORW/DME 108.8 COE Chan 25 N47°46.42' W116°49.24' at fld. 2320/19E. **HIWAS.**

DME portion unusable:

220°–240° byd 15 NM

280°–315° byd 15 NM blo 11,000'.

POST FALLS NDB (MHW) 347 LEN N47°44.57' W116°57.66' 053° 6.0 NM to fld.

ILS 110.7 I-COE Rwy 05 Class ID. Localizer unusable 25° left and right of course.

Figure 32. *Airport/facility directory excerpt.*

Useful load weights and moments

Baggage or 5th seat occupant

ARM 140

Weight	Moment/100
10	14
20	28
30	42
40	56
50	70
60	84
70	98
80	112
90	126
100	140
110	154
120	168
130	182
140	196
150	210
160	224
170	238
180	252
190	266
200	280
210	294
220	308
230	322
240	336
250	350
260	364
270	378

Occupants

Front seats ARM 85		Rear seats ARM 121	
Weight	Moment/100	Weight	Moment/100
120	102	120	145
130	110	130	157
140	119	140	169
150	128	150	182
160	136	160	194
170	144	170	206
180	153	180	218
190	162	190	230
200	170	200	242

Usable fuel

Main wing tanks ARM 75

Gallons	Weight	Moment/100
5	30	22
10	60	45
15	90	68
20	120	90
25	150	112
30	180	135
35	210	158
40	240	180
44	264	198

Auxiliary wing tanks ARM 94

Gallons	Weight	Moment/100
5	30	28
10	60	56
15	90	85
19	114	107

*Oil

Quarts	Weight	Moment/100
10	19	5

*Included in basic empty weight.

Empty weight~2,015
MOM/100~1,554
Moment limits vs weight
Moment limits are based on the following weight and center of gravity limit data (landing gear down).

Weight condition	Forward CG limit	AFT CG limit
2,950 lb (takeoff or landing)	82.1	84.7
2,525 lb	77.5	85.7
2,475 lb or less	77.0	85.7

Figure 33. *Airplane weight and balance tables.*

Moment limits vs weight (continued)						
Weight	Minimum Moment 100	Maximum Moment 100		Weight	Minimum Moment 100	Maximum Moment 100
2,100	1,617	1,800		2,500	1,932	2,143
2,110	1,625	1,808		2,510	1,942	2,151
2,120	1,632	1,817		2,520	1,953	2,160
2,130	1,640	1,825		2,530	1,963	2,168
2,140	1,648	1,834		2,540	1,974	2,176
2,150	1,656	1,843		2,550	1,984	2,184
2,160	1,663	1,851		2,560	1,995	2,192
2,170	1,671	1,860		2,570	2,005	2,200
2,180	1,679	1,868		2,580	2,016	2,208
2,190	1,686	1,877		2,590	2,026	2,216
2,200	1,694	1,885		2,600	2,037	2,224
2,210	1,702	1,894		2,610	2,048	2,232
2,220	1,709	1,903		2,620	2,058	2,239
2,230	1,717	1,911		2,630	2,069	2,247
2,240	1,725	1,920		2,640	2,080	2,255
2,250	1,733	1,928		2,650	2,090	2,263
2,260	1,740	1,937		2,660	2,101	2,271
2,270	1,748	1,945		2,670	2,112	2,279
2,280	1,756	1,954		2,680	2,123	2,287
2,290	1,763	1,963		2,690	2,133	2,295
2,300	1,771	1,971		2,700	2,144	2,303
2,310	1,779	1,980		2,710	2,155	2,311
2,320	1,786	1,988		2,720	2,166	2,319
2,330	1,794	1,997		2,730	2,177	2,326
2,340	1,802	2,005		2,740	2,188	2,334
2,350	1,810	2,014		2,750	2,199	2,342
2,360	1,817	2,023		2,760	2,210	2,350
2,370	1,825	2,031		2,770	2,221	2,358
2,380	1,833	2,040		2,780	2,232	2,366
2,390	1,840	2,048		2,790	2,243	2,374
2,400	1,848	2,057		2,800	2,254	2,381
2,410	1,856	2,065		2,810	2,265	2,389
2,420	1,863	2,074		2,820	2,276	2,397
2,430	1,871	2,083		2,830	2,287	2,405
2,440	1,879	2,091		2,840	2,298	2,413
2,450	1,887	2,100		2,850	2,309	2,421
2,460	1,894	2,108		2,860	2,320	2,428
2,470	1,902	2,117		2,870	2,332	2,436
2,480	1,911	2,125		2,880	2,343	2,444
2,490	1,921	2,134		2,890	2,354	2,452
				2,900	2,365	2,460
				2,910	2,377	2,468
				2,920	2,388	2,475
				2,930	2,399	2,483
				2,940	2,411	2,491
				2,950	2,422	2,499

Figure 34. *Airplane weight and balance tables.*

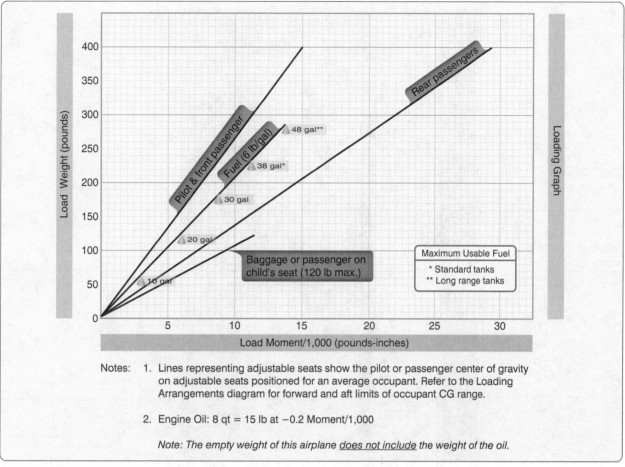

Notes: 1. Lines representing adjustable seats show the pilot or passenger center of gravity on adjustable seats positioned for an average occupant. Refer to the Loading Arrangements diagram for forward and aft limits of occupant CG range.

2. Engine Oil: 8 qt = 15 lb at −0.2 Moment/1,000

Note: The empty weight of this airplane does not include the weight of the oil.

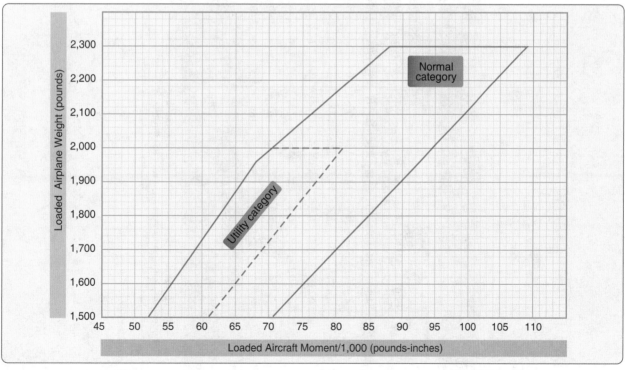

Figure 35. *Airplane weight and balance graphs.*

Cruise power settings
65% Maximum continuous power (or full throttle 2,800 pounds)

ISA −20 °C (−36 °F)

Press ALT. Feet	IOAT °F	IOAT °C	Engine speed RMP	MAN. press IN HG	Fuel flow per engine PSI	Fuel flow per engine GPH	TAS KTS	TAS MPH
SL	27	−3	2,450	20.7	6.6	11.5	147	169
2,000	19	−7	2,450	20.4	6.6	11.5	149	171
4,000	12	−11	2,450	20.1	6.6	11.5	152	175
6,000	5	−15	2,450	19.8	6.6	11.5	155	178
8,000	−2	−19	2,450	19.5	6.6	11.5	157	181
10,000	−8	−22	2,450	19.2	6.6	11.5	160	184
12,000	−15	−26	2,450	18.8	6.4	11.5	162	186
14,000	−22	−30	2,450	17.4	5.8	10.5	159	183
16,000	−29	−34	2,450	16.1	5.3	9.7	156	180

Standard day (ISA)

Press ALT. Feet	IOAT °F	IOAT °C	Engine speed RMP	MAN. press IN HG	Fuel flow per engine PSI	Fuel flow per engine GPH	TAS KTS	TAS MPH
SL	63	17	2,450	21.2	6.6	11.5	150	173
2,000	55	13	2,450	21.0	6.6	11.5	153	176
4,000	48	9	2,450	20.7	6.6	11.5	156	180
6,000	41	5	2,450	20.4	6.6	11.5	158	182
8,000	36	2	2,450	20.2	6.6	11.5	161	185
10,000	28	−2	2,450	19.9	6.6	11.5	163	188
12,000	21	−6	2,450	18.8	6.1	10.9	163	188
14,000	14	−10	2,450	17.4	5.6	10.1	160	184
16,000	7	−14	2,450	16.1	5.1	9.4	156	180

ISA +20 °C (+36 °F)

Press ALT. Feet	IOAT °F	IOAT °C	Engine speed RMP	MAN. press IN HG	Fuel flow per engine PSI	Fuel flow per engine GPH	TAS KTS	TAS MPH
SL	99	37	2,450	21.8	6.6	11.5	153	176
2,000	91	33	2,450	21.5	6.6	11.5	156	180
4,000	84	29	2,450	21.3	6.6	11.5	159	183
6,000	79	26	2,450	21.0	6.6	11.5	161	185
8,000	72	22	2,450	20.8	6.6	11.5	164	189
10,000	64	18	2,450	20.3	6.5	11.4	166	191
12,000	57	14	2,450	18.8	5.9	10.6	163	188
14,000	50	10	2,450	17.4	5.4	9.8	160	184
16,000	43	6	2,450	16.1	4.9	9.1	155	178

Note: 1. Full throttle manifold pressure settings are approximate.
2. Shaded area represents operation with full throttle.

Figure 36. *Airplane power setting table.*

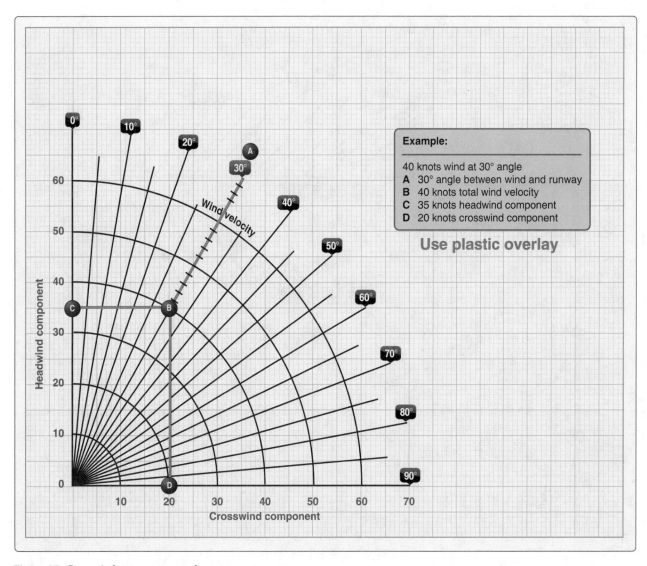

Figure 37. *Crosswind component graph.*

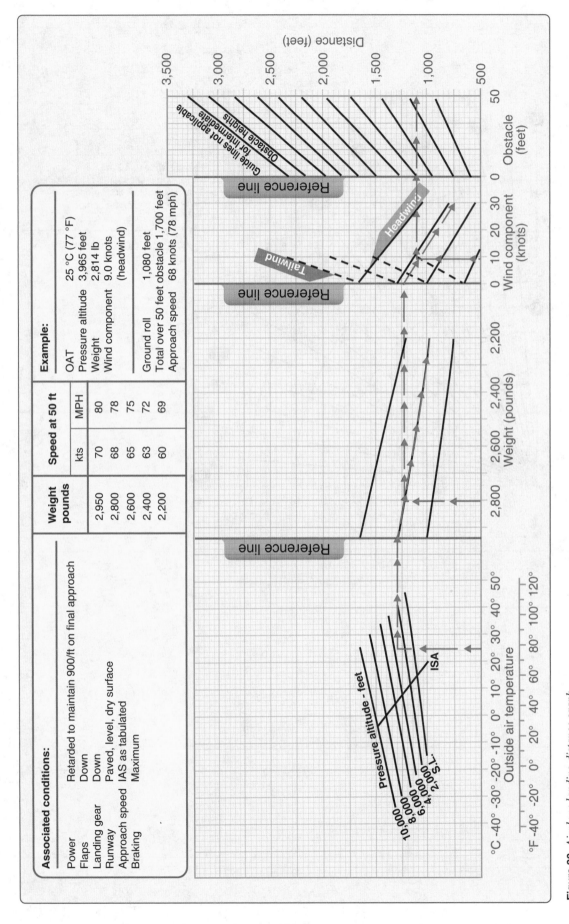

Figure 38. *Airplane landing distance graph.*

Landing distance

Flaps lowered to 40° – Power off
Hard surface runway – Zero wind

Gross weight lb	Approach speed, IAS, MPH	At sea level & 59 °F		At 2,500 feet & 50 °F		At 5,000 feet & 41 °F		At 7,500 feet & 32 °F	
		Ground roll	Total to clear 50 feet OBS	Ground roll	Total to clear 50 feet OBS	Ground roll	Total to clear 50 feet OBS	Ground roll	Total to clear 50 feet OBS
1,600	60	445	1,075	470	1,135	495	1,195	520	1,255

NOTE:
1. Decrease the distances shown by 10% for each 4 knots of headwind.
2. Increase the distance by 10% for each 60 °F temperature increase above standard.
3. For operation on a dry, grass runway, increase distance (both "ground roll" and "total to clear 50 feet obstacle") by 20% of the "total to clear 50 feet obstacle" figure.

Figure 39. *Airplane landing distance table.*

41

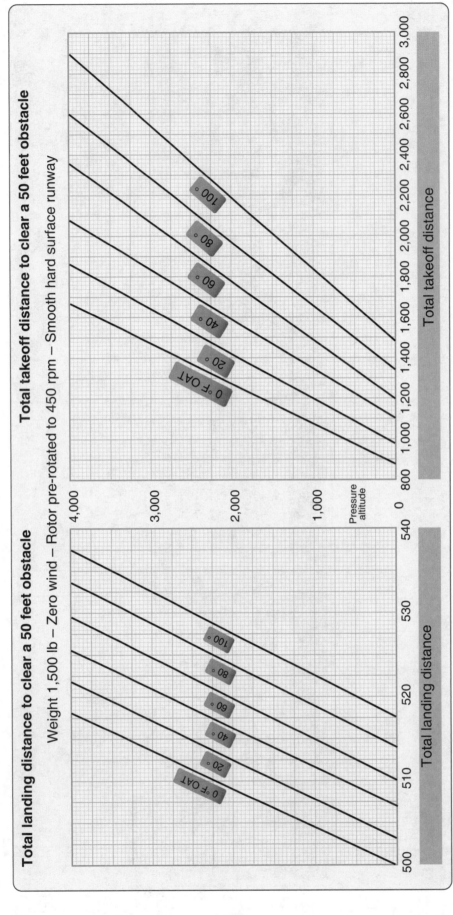

Figure 40. *Gyroplane takeoff and landing graphs.*

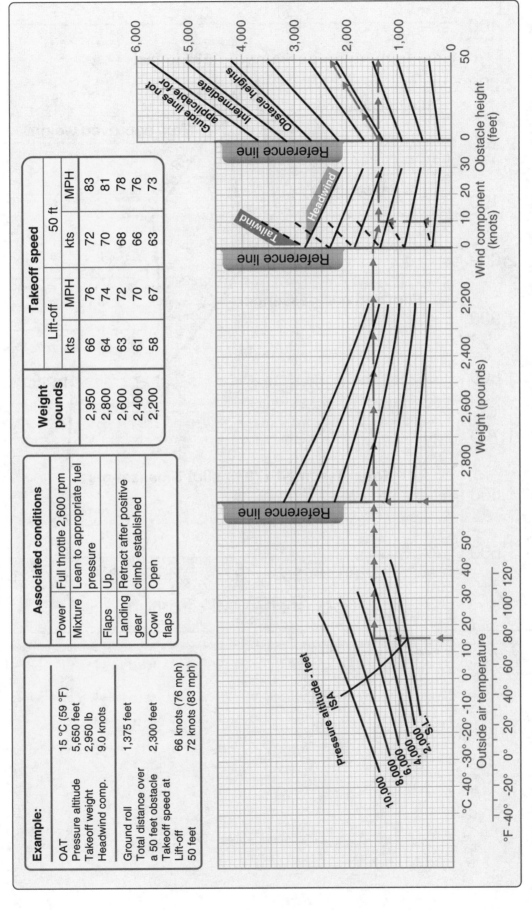

Figure 41. *Airplane takeoff distance graph.*

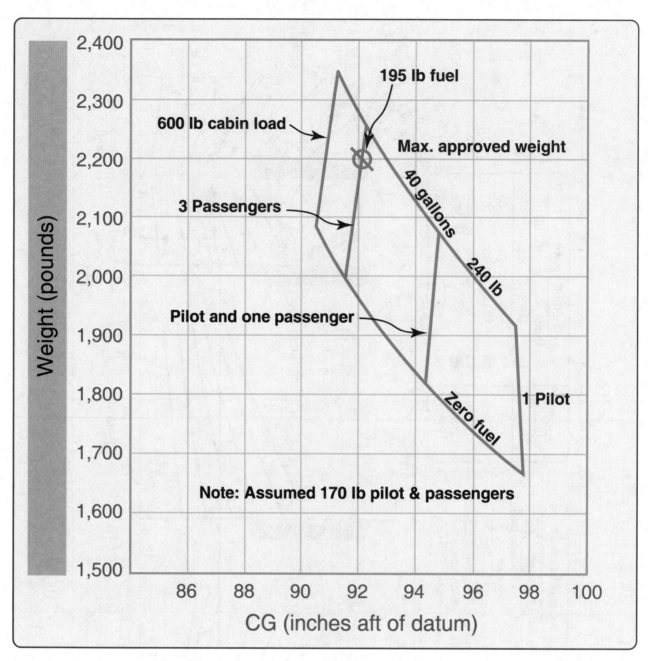

Figure 42. *Helicopter weight and balance graph.*

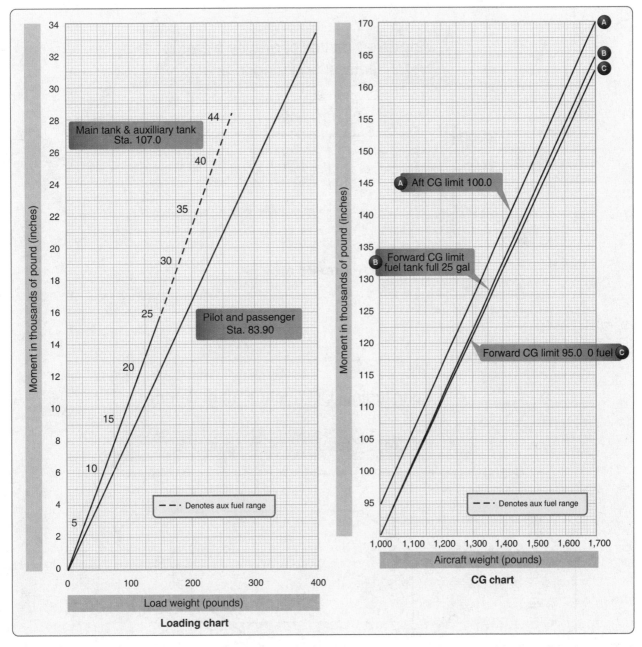

Figure 43. *Helicopter weight and balance graphs.*

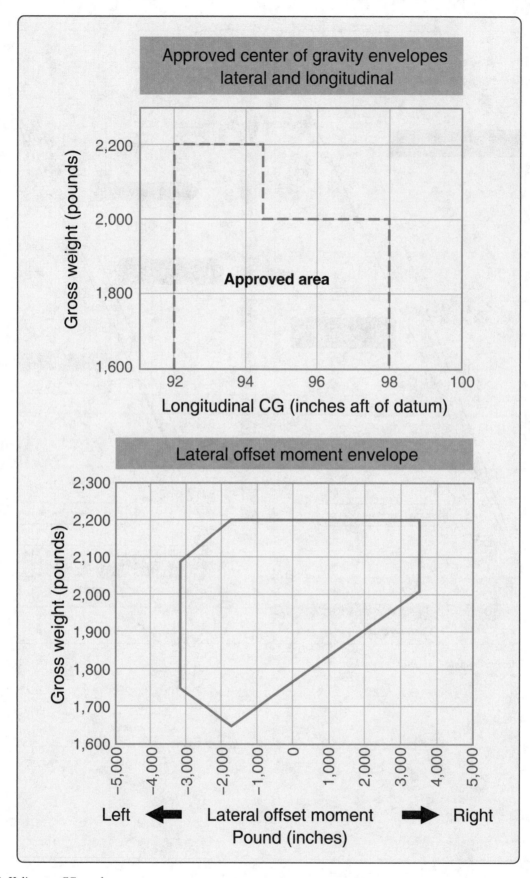

Figure 44. *Helicopter CG envelopes.*

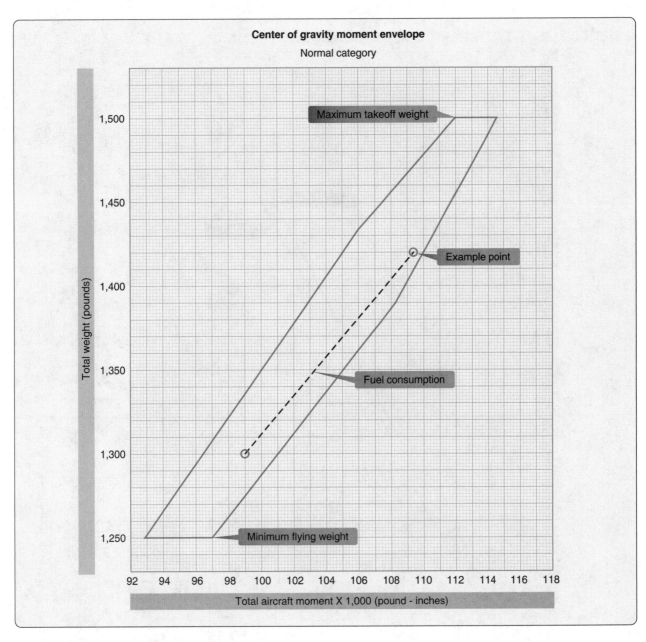

Figure 45. *Gyroplane weight and balance graph.*

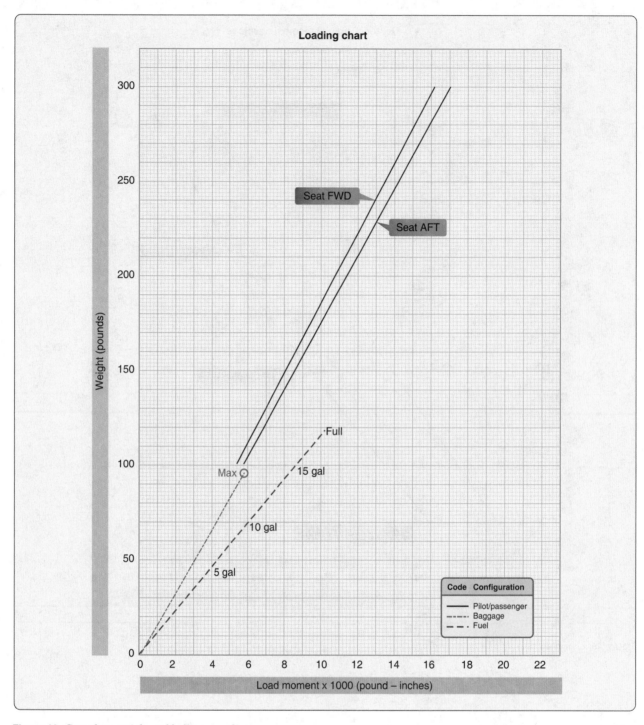

Figure 46. *Gyroplane weight and balance graph.*

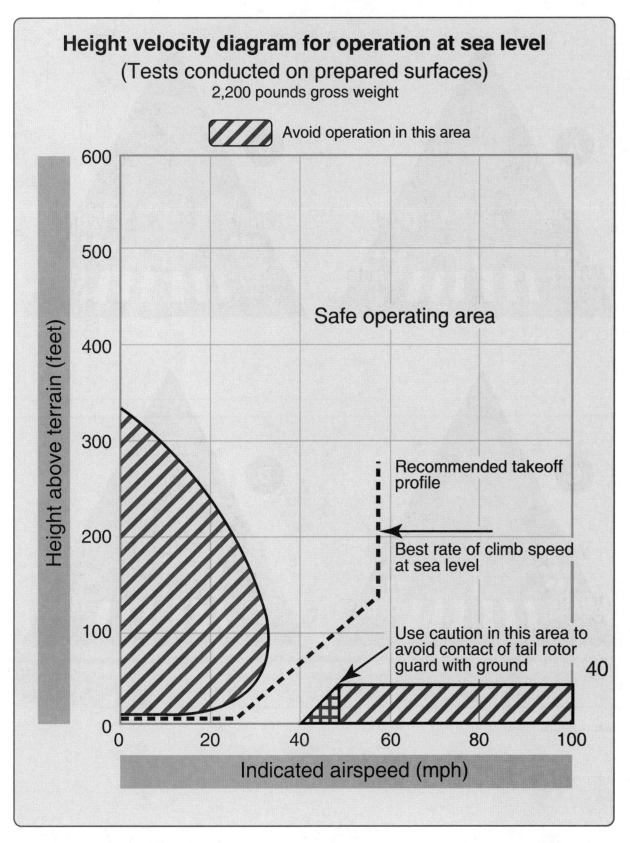

Figure 47. *Helicopter height velocity diagram.*

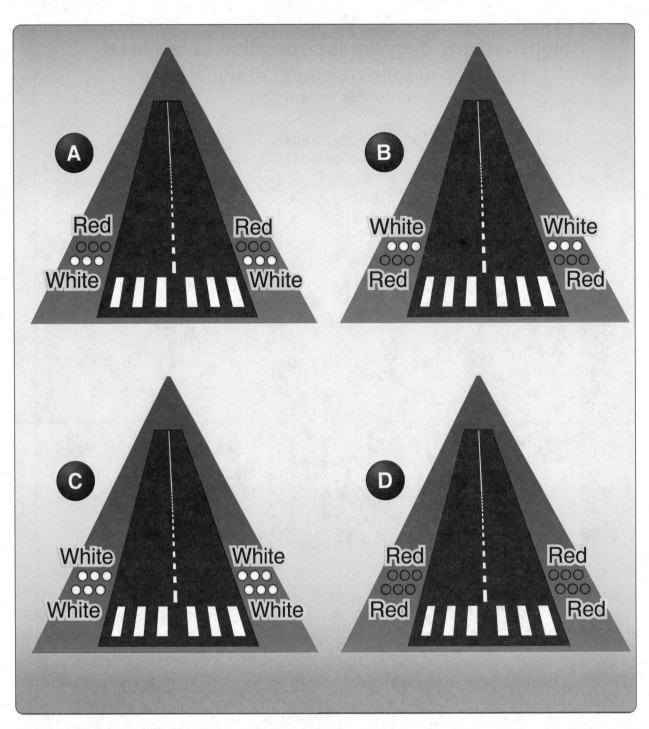

Figure 48. *VASI illustrations.*

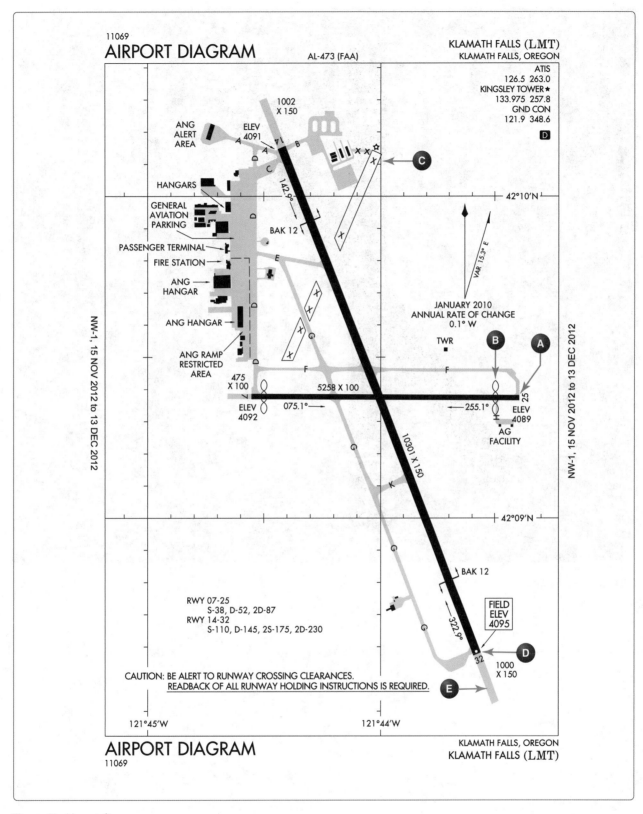

Figure 49. *Airport diagram.*

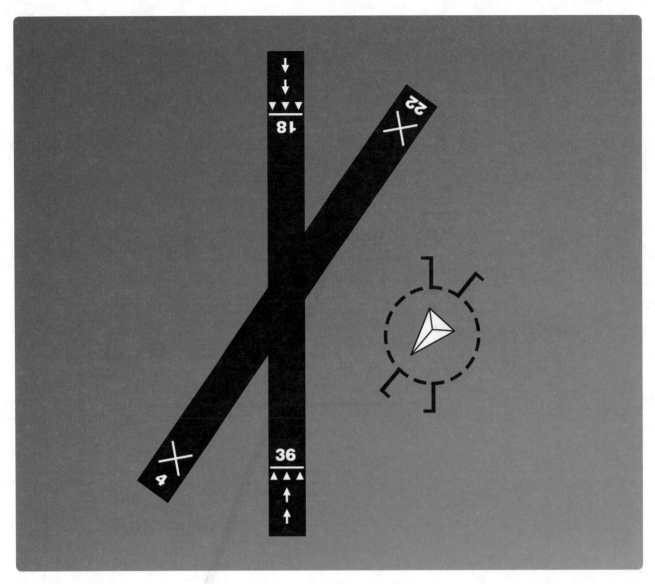

Figure 50. *Airport diagram.*

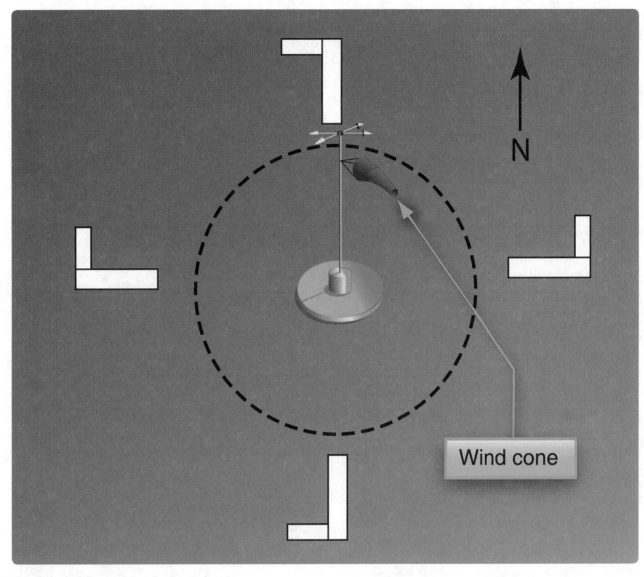

Figure 51. *Wind cone airport landing indicator.*

Form Approved OMB No. 2120-0026

U.S. DEPARTMENT OF TRANSPORTATION
FEDERAL AVIATION ADMINISTRATION

FLIGHT PLAN

(FAA USE ONLY) ☐ PILOT BRIEFING ☐ VNR ☐ STOPOVER

TIME STARTED

SPECIALIST INITIALS

1. TYPE	2. AIRCRAFT IDENTIFICATION	3. AIRCRAFT TYPE/ SPECIAL EQUIPMENT	4. TRUE AIRSPEED	5. DEPARTURE POINT	6. DEPARTURE TIME		7. CRUISING ALTITUDE
VFR					PROPOSED (Z)	ACTUAL (Z)	
IFR							
DVFR			KTS				

8. ROUTE OF FLIGHT

9. DESTINATION (Name of airport and city)	10. EST. TIME ENROUTE		11. REMARKS
	HOURS	MINUTES	

12. FUEL ON BOARD		13. ALTERNATE AIRPORT(S)	14. PILOT'S NAME, ADDRESS & TELEPHONE NUMBER & AIRCRAFT HOME BASE	15. NUMBER ABOARD
HOURS	MINUTES		17. DESTINATION CONTACT/TELEPHONE (OPTIONAL)	

16. COLOR OF AIRCRAFT

CIVIL AIRCRAFT PILOTS. 14 CFR Part 91 requires you file an IFR flight plan to operate under instrument flight rules in controlled airspace. Failure to file could result in a civil penalty not to exceed $1,000 for each violation (Section 901 of the Federal Aviation Act of 1958, as amended). Filing of a VFR flight plan is recommended as a good operating practice. See also Part 99 for requirements concerning DVFR flight plans.

FAA Form 7233-1 (8-82)

CLOSE VFR FLIGHT PLAN WITH ____________ FSS ON ARRIVAL

Figure 52. *Flight plan form.*

LINCOLN (LNK) 4 NW UTC−6(−5DT) N40°51.05′ W96°45.55′   OMAHA

1219 B S4 **FUEL** 100LL, JET A TPA—See Remarks ARFF Index—See Remarks   H–5C, L–10I

NOTAM FILE LNK   IAP, AD

RWY 18–36: H12901X200 (ASPH–CONC–GRVD) S–100, D–200,
     2S–175, 2D–400 HIRL
   RWY 18: MALSR. PAPI(P4L)—GA 3.0° TCH 55′. Rgt tfc. 0.4%
     down.
   RWY 36: MALSR. PAPI(P4L)—GA 3.0° TCH 57′.

RWY 14–32: H8649X150 (ASPH–CONC–GRVD) S–80, D–170,
     2S–175, 2D–280 MIRL
   RWY 14: REIL. VASI(V4L)—GA 3.0° TCH 48′. Thld dsplcd 363′.
   RWY 32: VASI(V4L)—GA 3.0° TCH 50′. Thld dsplcd 470′.
     Pole. 0.3% up.

RWY 17–35: H5800X100 (ASPH–CONC–AFSC) S–49, D–60
     HIRL 0.8% up S
   RWY 17: REIL. PAPI(P4L)—GA 3.0° TCH 44′.
   RWY 35: ODALS. PAPI(P4L)—GA 3.0° TCH 30′. Rgt tfc.

RUNWAY DECLARED DISTANCE INFORMATION

RWY	TORA	TODA	ASDA	LDA
RWY 14:	TORA–8649	TODA–8649	ASDA–8649	LDA–8286
RWY 17:	TORA–5800	TODA–5800	ASDA–5400	LDA–5400
RWY 18:	TORA–12901	TODA–12901	ASDA–12901	LDA–12901
RWY 32:	TORA–8649	TODA–8649	ASDA–8286	LDA–7816
RWY 35:	TORA–5800	TODA–5800	ASDA–5800	LDA–5800
RWY 36:	TORA–12901	TODA–12901	ASDA–12901	LDA–12901

AIRPORT REMARKS: Attended continuously. Birds invof arpt. Rwy 18 designated calm wind rwy. Rwy 32 apch holdline
on South A twy. TPA–2219 (1000), heavy military jet 3000 (1781). Class I, ARFF Index B. ARFF Index C level
equipment provided. Rwy 18–36 touchdown and rollout rwy visual range avbl. When twr clsd MIRL Rwy 14–32
preset on low ints, HIRL Rwy 18–36 and Rwy 17–35 preset on med ints, ODALS Rwy 35 operate continuously on
med ints, MALSR Rwy 18 and Rwy 36 operate continuously and REIL Rwy 14 and Rwy 17 operate continuously
on low ints. VASI Rwy 14 and Rwy 32, PAPI Rwy 17, Rwy 35, Rwy 18 and Rwy 36 on continuously.

WEATHER DATA SOURCES: ASOS (402) 474–9214. LLWAS

COMMUNICATIONS: CTAF 118.5   **ATIS** 118.05   **UNICOM** 122.95

   RCO 122.65 (COLUMBUS RADIO)

Ⓡ **APP/DEP CON** 124.0 (180°–359°) 124.8 (360°–179°)

   TOWER 118.5 125.7 (1130–0600Z‡)   **GND CON** 121.9   **CLNC DEL** 120.7

AIRSPACE: CLASS C svc 1130–0600Z‡ ctc **APP CON** other times CLASS E.

RADIO AIDS TO NAVIGATION: NOTAM FILE LNK.

   (H) VORTACW 116.1   LNK   Chan 108   N40°55.43′ W96°44.52′   181° 4.4 NM to fld. 1370/9E

   POTTS NDB (MHW/LOM) 385   LN   N40°44.83′ W96°45.75′   355° 6.2 NM to fld. Unmonitored when twr clsd.

   ILS 111.1   I–OCZ   Rwy 18.   Class IB   OM unmonitored.

   ILS 109.9   I–LNK   Rwy 36   Class IA   LOM POTTS NDB. MM unmonitored. LOM unmonitored when twr
     clsd.

COMM/NAV/WEATHER REMARKS: Emerg frequency 121.5 not available at twr.

LOUP CITY MUNI (ØF4) 1 NW UTC−6(−5DT) N41°17.20′ W98°59.41′   OMAHA

2071 B **FUEL** 100LL NOTAM FILE OLU   L–10H, 12H

RWY 16–34: H3200X60 (CONC) S–12.5 MIRL
   RWY 34: Trees.

RWY 04–22: 2040X100 (TURF)
   RWY 04: Tree.   RWY 22: Road.

AIRPORT REMARKS: Unattended. For svc call 308–745–1344/1244/0664.

COMMUNICATIONS: CTAF 122.9

RADIO AIDS TO NAVIGATION: NOTAM FILE OLU.

   WOLBACH (H) VORTAC 114.8   OBH   Chan 95   N41°22.54′ W98°21.22′   253° 29.3 NM to fld. 2010/7E.

MARTIN FLD (See SO SIOUX CITY)

Figure 53. *Airport/facility directory excerpt.*

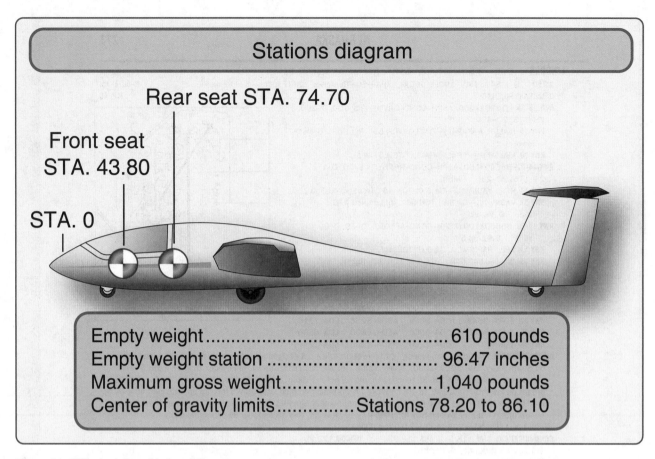

Figure 54. *Glider weight and balance diagram.*

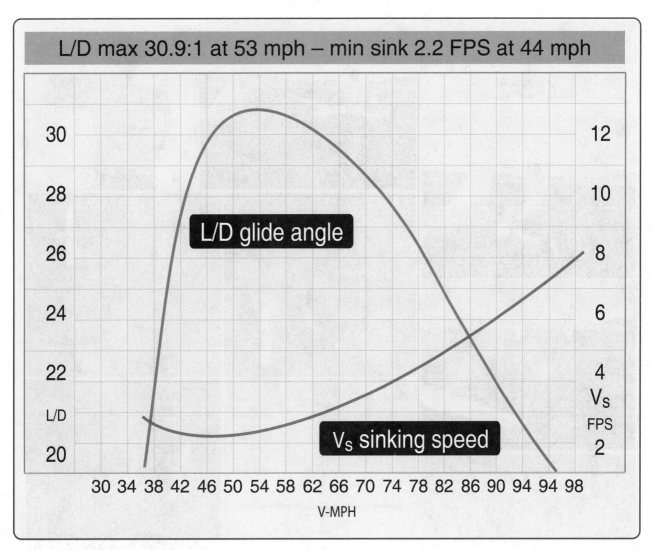

Figure 55. *Glider performance graph.*

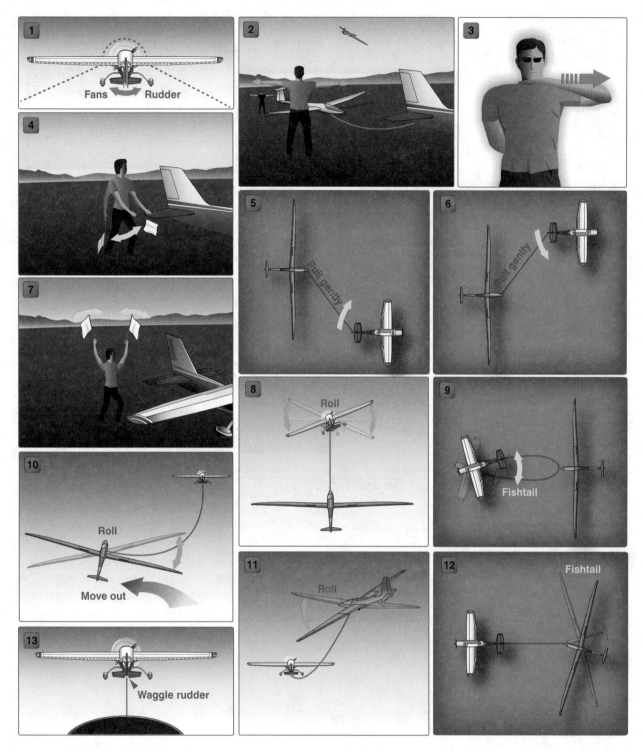

Figure 56. *Standard soaring signals.*

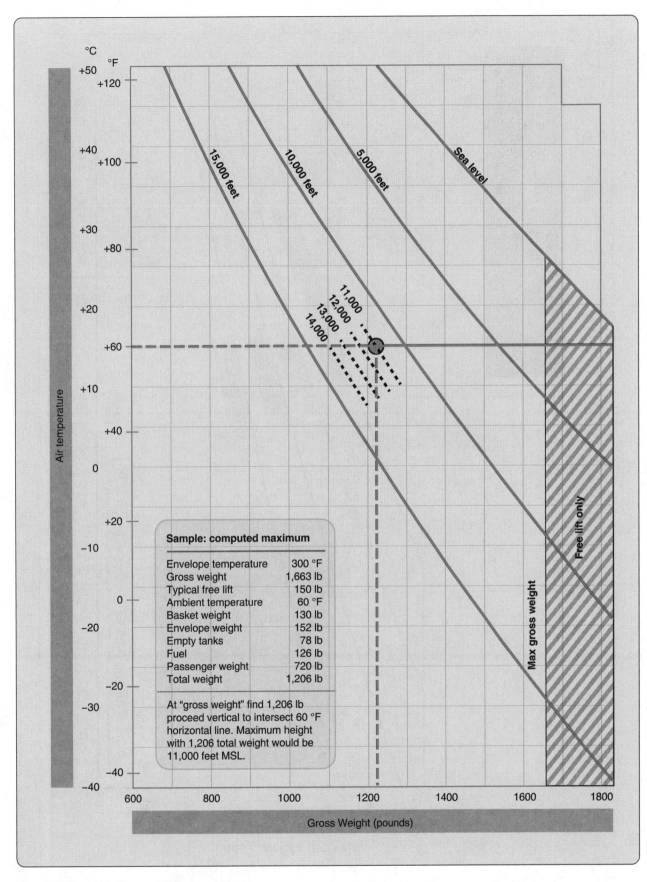

Figure 57. *Hot air balloon performance graph.*

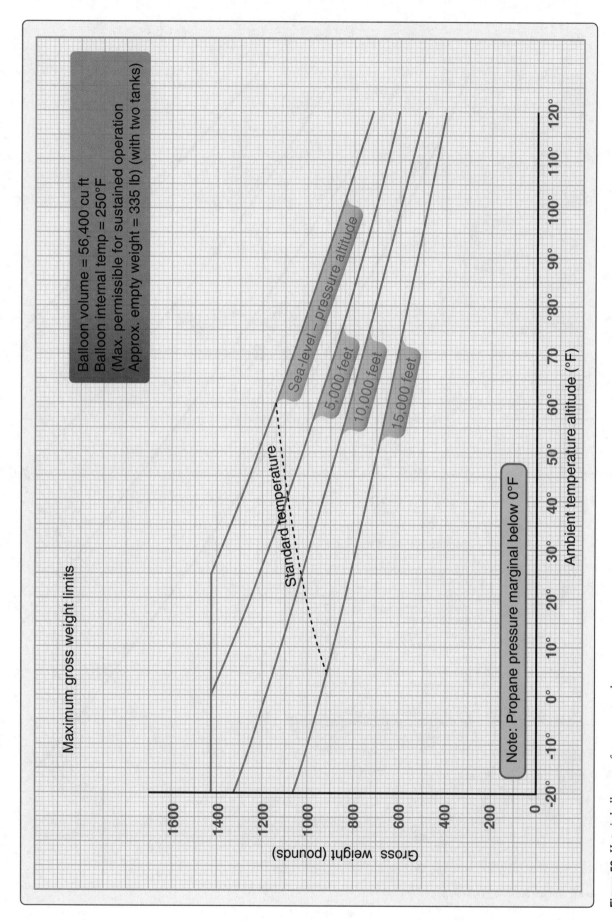

Figure 58. *Hot air balloon performance graph.*

60

For	N	30	60	E	120	150
Steer	0	27	56	85	116	148
For	S	210	240	W	300	330
Steer	181	214	244	274	303	332

Figure 59. *Compass card.*

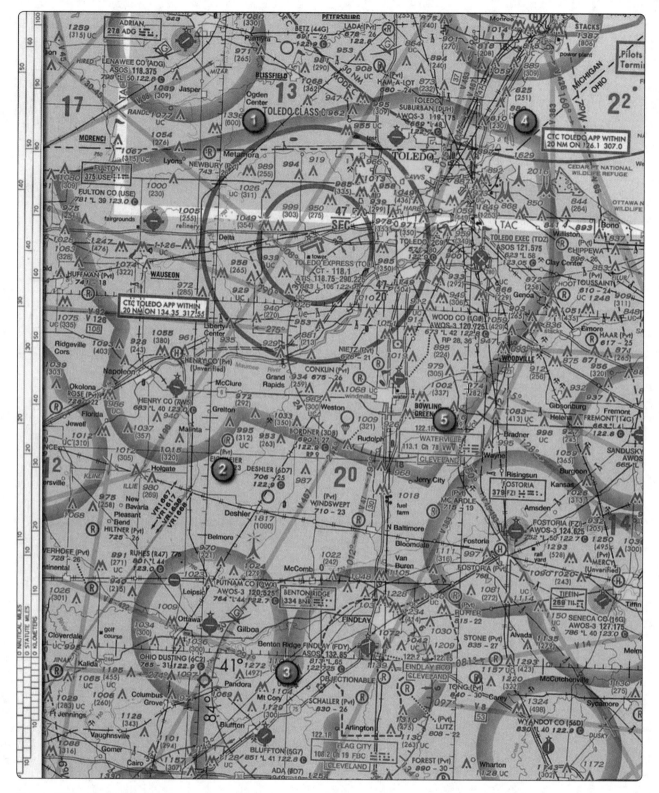

Figure 60. *Sectional chart excerpt.*

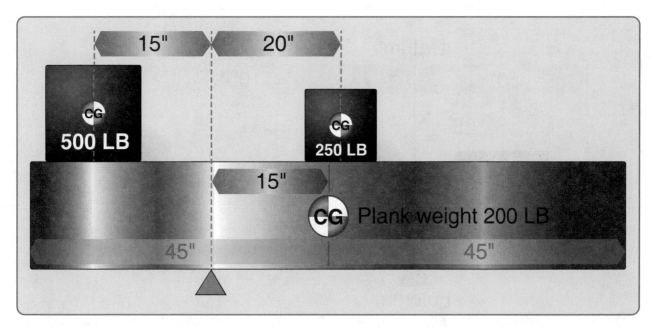

Figure 61. *Weight and balance diagram.*

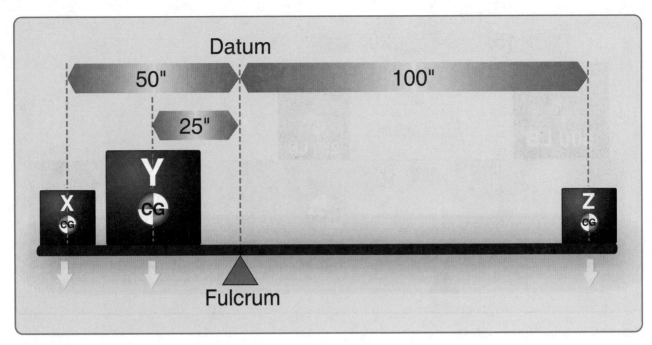

Figure 62. *Weight and balance diagram.*

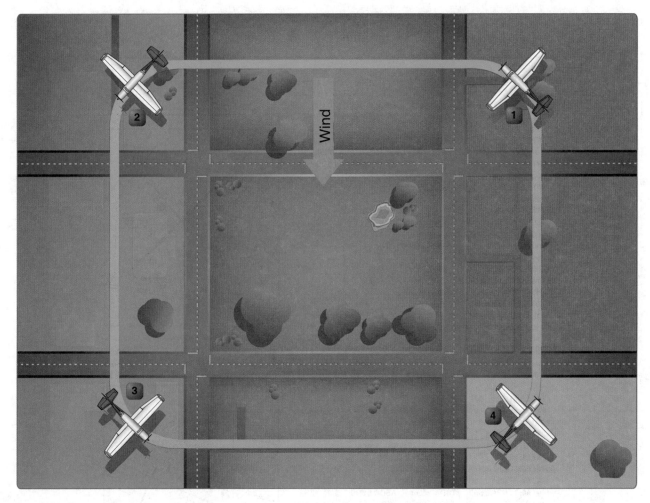

Figure 63. *Rectangular course.*

TOLEDO

TOLEDO EXECUTIVE (TDZ)   6 SE   UTC−5(−4DT)   N41°33.90′ W83°28.93′   **DETROIT**
623   B   S4   **FUEL** 100LL, JET A   OX 1, 3   NOTAM FILE TDZ   H−10G, L−28J
RWY 14–32: H5829X100 (ASPH–GRVD)   S−63, D−85, 2S−107   MIRL   **IAP**
  RWY 14: REIL. PAPI(P4L)—GA 3.0° TCH 34′. Thld dsplcd 225′.
  Tower.
  RWY 32: VASI(V4L)—GA 3.0° TCH 43′. Thld dsplcd 351′. Road.
RWY 04–22: H3799X75 (ASPH)   S−63, D−85, 2S−107   MIRL
  RWY 04: PAPI(P4L)—GA 3.5° TCH 35′. Thld dsplcd 100′.
  Road.
  RWY 22: REIL. PAPI(P4L)—GA 3.0° TCH 25′. Thld dsplcd 380′.
  Railroad.
AIRPORT REMARKS: Attended Mon–Fri continuously, Sat–Sun
  1300–0100Z‡. Parallel twy Rwy 04–22 and Rwy 14–32 35′ wide.
  Seagulls on and invof arpt. Ldg fee. ACTIVATE MIRL Rwy 04–22
  and Rwy 14–32, REIL and PAPI Rwy 04, Rwy 22, Rwy 14 and VASI
  Rwy 32—CTAF.
WEATHER DATA SOURCES: ASOS 121.575 (419) 838–5034.
COMMUNICATIONS: CTAF/UNICOM 123.05
Ⓡ **APP/DEP CON** 126.1   **CLNC DEL** 125.6
RADIO AIDS TO NAVIGATION: NOTAM FILE CLE.
  WATERVILLE (L) VOR/DME 113.1   VWV   Chan 78   N41°27.09′
  W83°38.32′   048° 9.8 NM to fld. 664/2W.

TOLEDO EXPRESS (TOL)   10 W   UTC−5(−4DT)   N41°35.21′ W83°48.47′   **DETROIT**
683   B   S4   **FUEL** 100LL, JET A   OX 3   LRA   Class I, ARFF Index B   NOTAM FILE TOL   H−10G, L−28J
RWY 07–25: H10599X150 (ASPH–GRVD)   S−100, D−174, 2S−175, 2D−300, 2D/2D2−550   **IAP, AD**
  HIRL   CL
  RWY 07: ALSF2. TDZL. Trees.
  RWY 25: MALSR. VASI(V4L)—GA 3.0° TCH 51′. Trees.   0.3% up.
RWY 16–34: H5599X150 (ASPH–GRVD)   S−100, D−174, 2S−175,
  2D−300   MIRL
  RWY 16: REIL. PAPI(P4L)—GA 3.0° TCH 48′. Trees.
  RWY 34: REIL.
RUNWAY DECLARED DISTANCE INFORMATION

	TORA	TODA	ASDA	LDA
RWY 07:	10599	10599	10599	10599
RWY 16:	5599	5599	5599	5599
RWY 25:	10599	10599	10599	10599
RWY 34:	5599	5599	5599	5599

ARRESTING GEAR/SYSTEM
  RWY 07 ←BAK−12   BAK−12 →RWY 25
AIRPORT REMARKS: Attended continuously. Fuel and svc avbl
  1300–0500Z‡. Birds on and invof arpt. Twy A west of Rwy 16 and
  the ramp between Twy B9 and B13 not visible from twr. Twy D
  intersection of Twy D1, heavy acft use minimal power to reduce
  foreign object damage on Air National Guard ramp. Customs:
  Sat–Sun req must be made prior to 2200Z‡ on Fri, phone 419–259–6424.
WEATHER DATA SOURCES: ASOS (419) 865–8351.
COMMUNICATIONS: ATIS 118.75   UNICOM 122.95
Ⓡ **APP/DEP CON** 126.1 (360°–179°) 134.35 (180°–359°) 123.975
  TOWER 118.1   **GND CON** 121.9   **CLNC DEL** 121.75
AIRSPACE: CLASS C svc continuous ctc **APP CON**
RADIO AIDS TO NAVIGATION: NOTAM FILE CLE.
  WATERVILLE (L) VOR/DME 113.1   VWV   Chan 78   N41°27.09′ W83°38.32′   319° 11.1 NM to fld. 664/2W.
  TOPHR NDB (LOM) 219   TO   N41°33.21′ W83°55.27′   074° 5.5 NM to fld. Unmonitored. NOTAM FILE TOL.
  ILS 109.7   I−TOL   Rwy 07.   Class IE.   LOM TOPHR NDB.
  ILS 108.7   I−BQE   Rwy 25.   Class IA.   LOC unusable 0.4 NM inbound. ILS unmonitored when twr clsd.
  ASR

SEAGATE HELISTOP (6T2)   00 N   UTC−5(−4DT)   N41°39.25′ W83°31.88′   **DETROIT**
650   NOTAM FILE CLE
HELIPAD H1: H50X50 (CONC)
HELIPORT REMARKS: Unattended. ACTIVATE orange perimeter lgts—CTAF. Helipad H1 NSTD 1–box (2 VASIS). Helipad
  H1 not marked with ''H.'' Helipad H1 perimeter lgts.
COMMUNICATIONS: CTAF/UNICOM 123.05

Figure 64. *Airport/facility excerpt.*

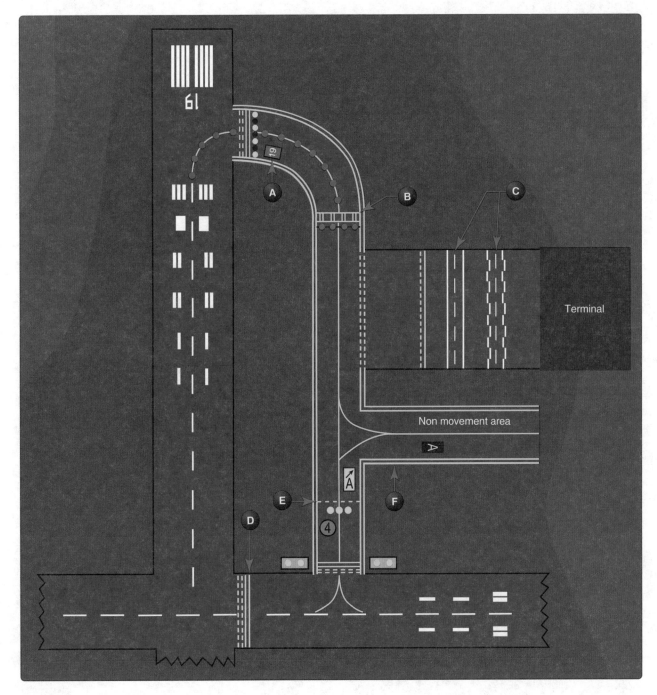

Figure 65. *Airport markings.*

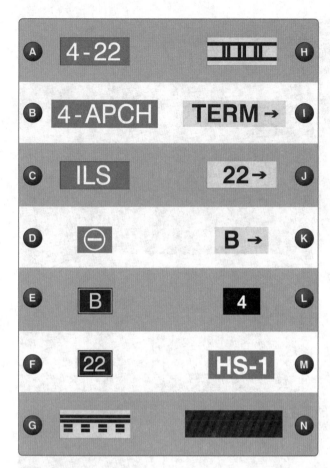

Figure 66. *U.S. airport signs.*

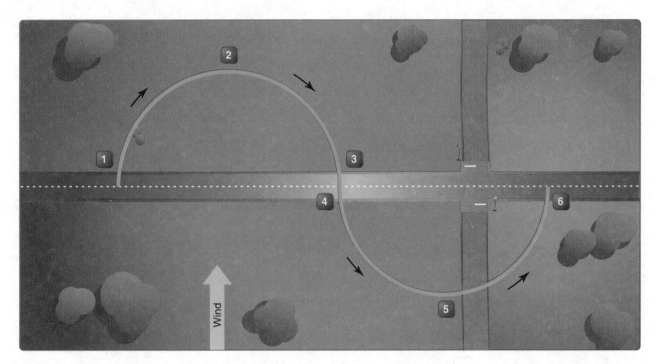

Figure 67. *S-turn diagram.*

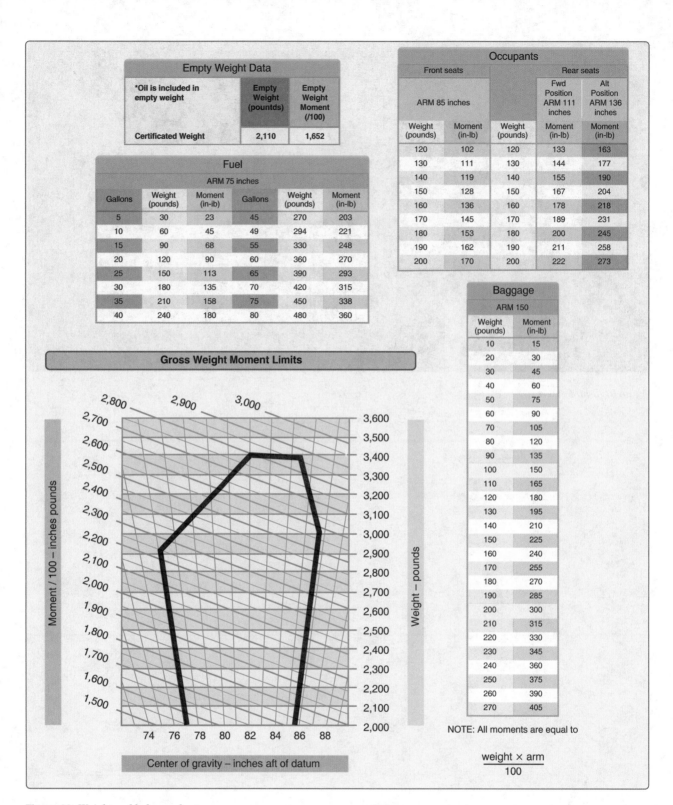

Empty Weight Data

*Oil is included in empty weight	Empty Weight (pountds)	Empty Weight Moment (/100)
Certificated Weight	2,110	1,652

Fuel

ARM 75 inches

Gallons	Weight (pounds)	Moment (in-ib)	Gallons	Weight (pounds)	Moment (in-lb)
5	30	23	45	270	203
10	60	45	49	294	221
15	90	68	55	330	248
20	120	90	60	360	270
25	150	113	65	390	293
30	180	135	70	420	315
35	210	158	75	450	338
40	240	180	80	480	360

Occupants

Front seats		Rear seats		
ARM 85 inches			Fwd Position ARM 111 inches	Alt Position ARM 136 inches
Weight (pounds)	Moment (in-lb)	Weight (pounds)	Moment (in-lb)	Moment (in-lb)
120	102	120	133	163
130	111	130	144	177
140	119	140	155	190
150	128	150	167	204
160	136	160	178	218
170	145	170	189	231
180	153	180	200	245
190	162	190	211	258
200	170	200	222	273

Baggage

ARM 150

Weight (pounds)	Moment (in-lb)
10	15
20	30
30	45
40	60
50	75
60	90
70	105
80	120
90	135
100	150
110	165
120	180
130	195
140	210
150	225
160	240
170	255
180	270
190	285
200	300
210	315
220	330
230	345
240	360
250	375
260	390
270	405

NOTE: All moments are equal to

$$\frac{weight \times arm}{100}$$

Gross Weight Moment Limits

Moment / 100 – inches pounds

Weight – pounds

Center of gravity – inches aft of datum

Figure 68. *Weight and balance chart.*

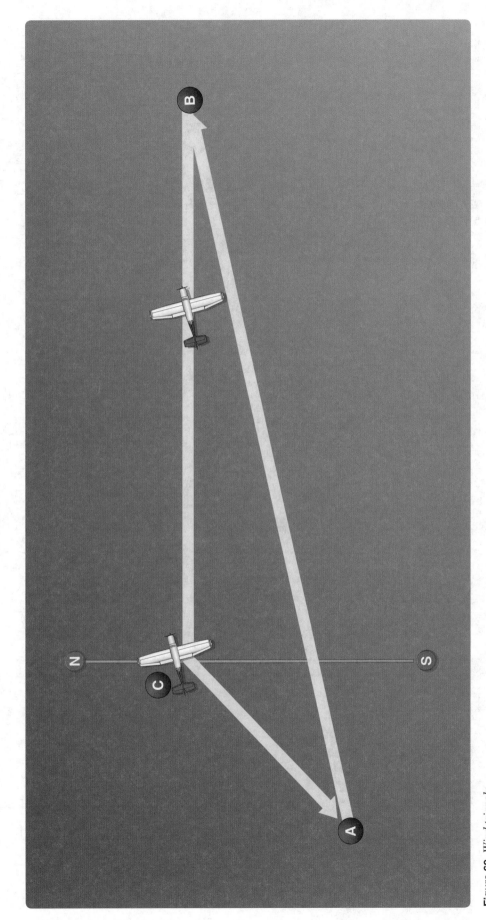

Figure 69. *Wind triangle.*

71

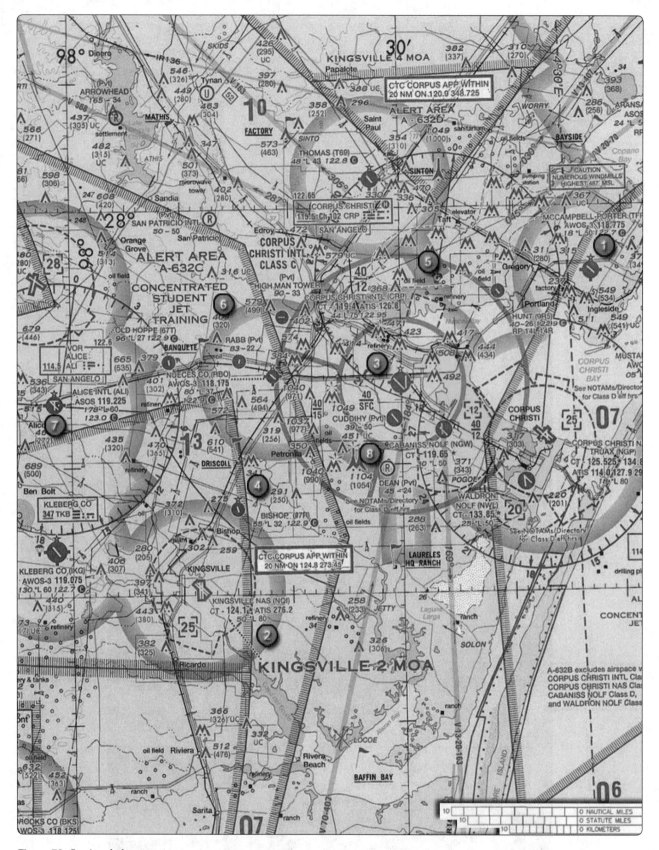

Figure 70. *Sectional chart excerpt.*

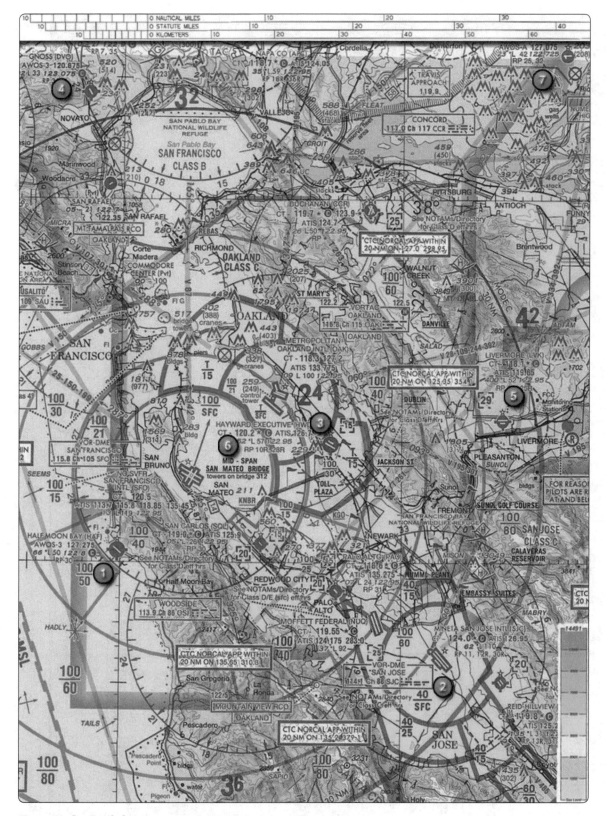

Figure 71. *Sectional chart excerpt.*

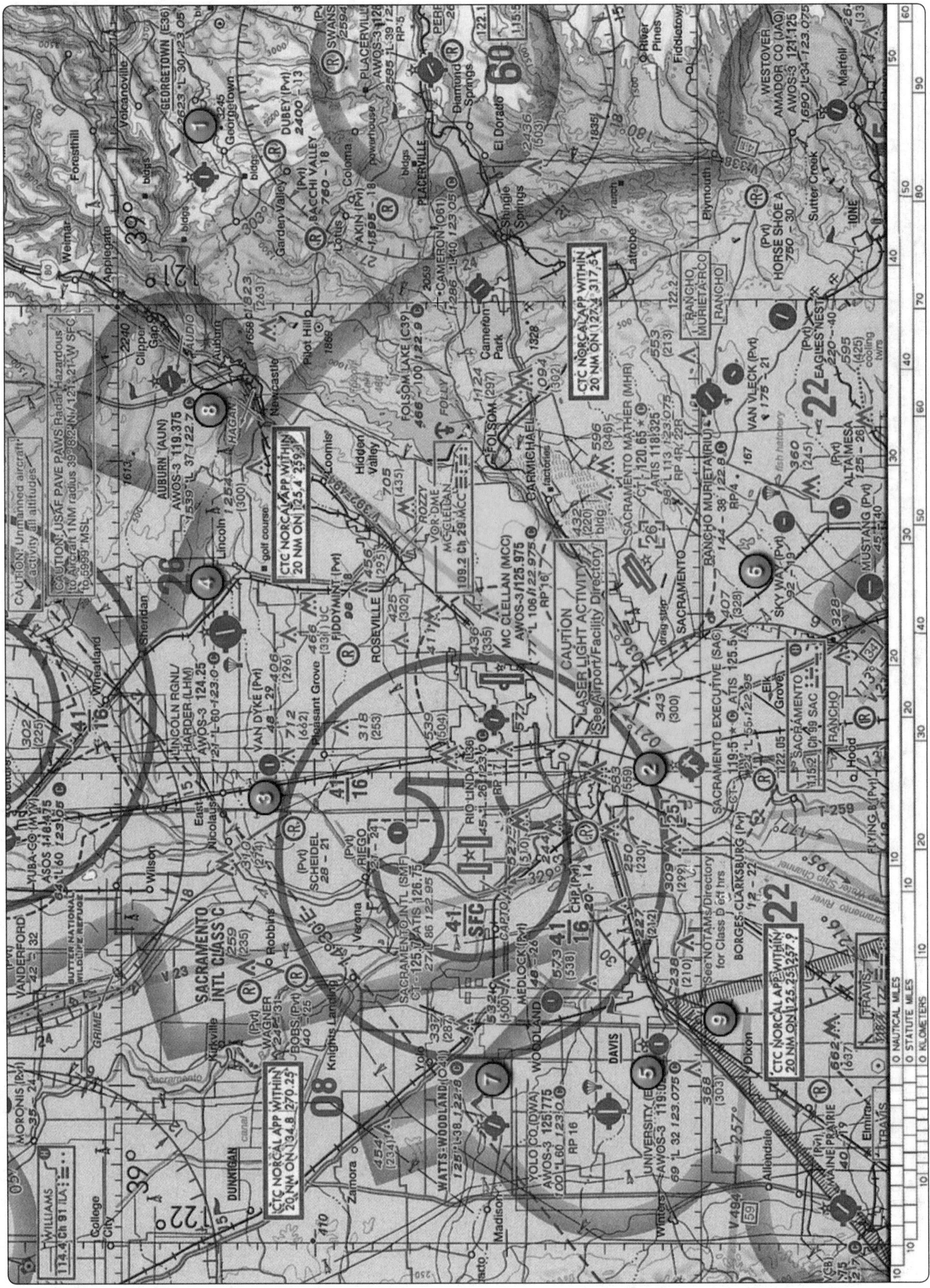

Figure 72. *Sectional chart excerpt.*

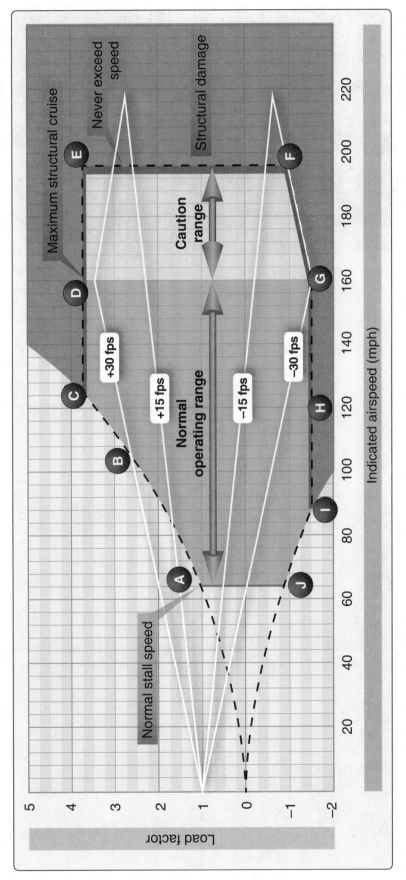

Figure 73. *Velocity vs. G-loads.*

75

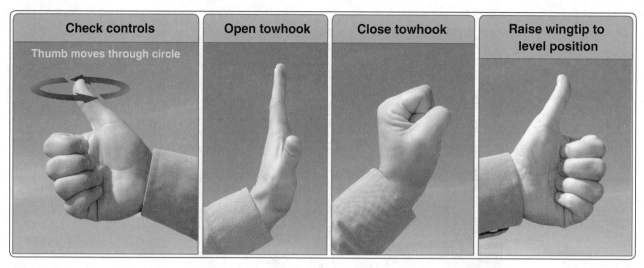

Figure 74. *Glider hand signals.*

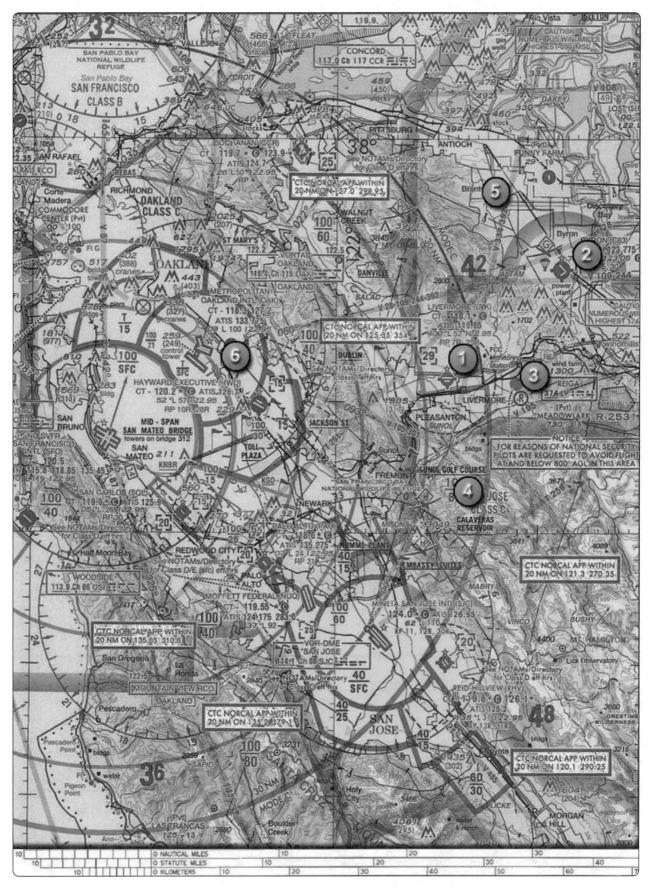

Figure 75. *Sectional chart excerpt.*

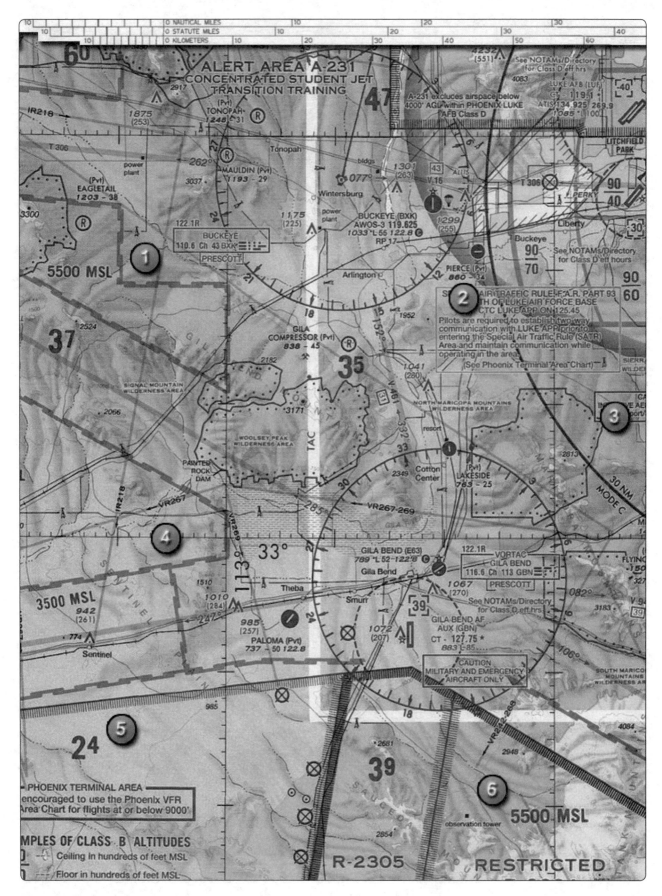

Figure 76. *Sectional chart excerpt.*

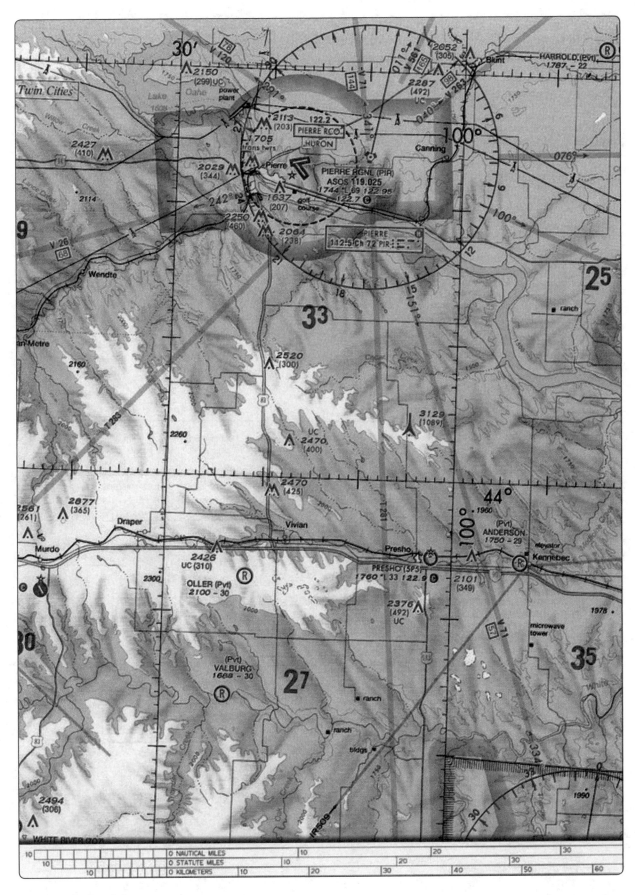

Figure 77. *Sectional chart excerpt.*

79

PIERRE RGNL (PIR) 3 E UTC −6(−5DT) N44°22.96′ W100°17.16′   OMAHA
1744 B S4 **FUEL** 100LL, JET A OX 1, 2, 3, 4 Class I, ARFF Index A NOTAM FILE PIR   H−2I, L−12H
RWY 13−31: H6900X100 (ASPH−GRVD) S−91, D−108, 2S−137, 2D−168 HIRL   IAP
  RWY 13: REIL. PAPI(P4L)—GA 3.0 ° TCH 52′.
  RWY 31: MALSR. PAPI(P4L)—GA 3.0 ° TCH 52′.
RWY 07−25: H6881X150 (ASPH−GRVD) S−91, D−114, 2S−145,
  2D−180 HIRL 0.6% up W
  RWY 07: REIL. PAPI(P4L)—GA 3.0 ° TCH 47′. Tank.
  RWY 25: REIL. PAPI(P4L)—GA 3.0 ° TCH 54′.
RUNWAY DECLARED DISTANCE INFORMATION
  RWY 07: TORA−6881 TODA−6881 ASDA−6830 LDA−6830
  RWY 13: TORA−6900 TODA−6900 ASDA−6900 LDA−6900
  RWY 25: TORA−6881 TODA−6881 ASDA−6881 LDA−6881
  RWY 31: TORA−6900 TODA−6900 ASDA−6900 LDA−6900
AIRPORT REMARKS: Attended Mon-Fri 1100-0600Z‡, Sat-Sun
  1100-0400Z‡. For attendant other times call
  605-224-9000/8621. Arpt conditions unmonitored during
  0530-1000Z‡. Numerous non-radio acft operating in area. Birds
  on and invof arpt and within a 25 NM radius. No line of sight
  between rwy ends of Rwy 07-25. ARFF provided for part 121 air
  carrier ops only. 48 hr PPR for unscheduled acr ops involving acft
  designed for 31 or more passenger seats call 605-773-7447 or
  605-773-7405. Taxiway C is 50′ wide and restricted to acft 75,000 pounds or less. ACTIVATE HIRL Rwy 13-31
  and Rwy 07-25, MALSR Rwy 31, REIL Rwy 07, Rwy 13 and Rwy 25, PAPI Rwy 07, Rwy 25, Rwy 13 and Rwy
  31—CTAF 122.7. NOTE: See Special Notices Section—
  Aerobatic Practice Areas.
WEATHER DATA SOURCES: ASOS 119.025 (605) 224-6087. HIWAS 112.5 PIR.
COMMUNICATIONS: CTAF 122.7   **UNICOM** 122.95
  RCO 122.2 (HURON RADIO)
Ⓡ **MINNEAPOLIS CENTER APP/DEP CON** 125.1
RADIO AIDS TO NAVIGATION: NOTAM FILE PIR.
  (L) VORTACW 112.5 PIR Chan 72 N44°23.67′ W100°09.77′ 251° 5.3 NM to fld. 1789/11E. **HIWAS.**
  ILS/DME 111.9 I-PIR Chan 56 Rwy 31. Class IA ILS GS unusable for coupled apch blo 2,255 ′. GS
    unusable blo 2135′.

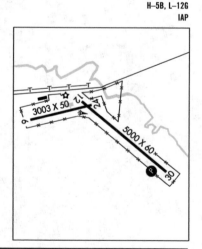

PINE RIDGE (IEN) 2 E UTC −7(−6DT) N43°01.35′ W102°30.66′   CHEYENNE
3333 B NOTAM FILE IEN   H−5B, L−12G
RWY 12−30: H5000X60 (ASPH) S−12 MIRL 0.7% up SE   IAP
  RWY 12: P-line.
  RWY 30: PAPI(P2L)—GA 3.0 ° TCH 26′. Fence.
RWY 06−24: H3003X50 (ASPH) S−12 0.7% up NE
  RWY 24: Fence.
AIRPORT REMARKS: Unattended. Rwy 06-24 CLOSED indef. MIRL Rwy
  12-30 and PAPI Rwy 30 opr dusk-0530Z‡, after 0530Z‡
  ACTIVATE—CTAF.
WEATHER DATA SOURCES: ASOS 126.775 (605) 867-1584.
COMMUNICATIONS: CTAF 122.9
  DENVER CENTER APP/DEP CON 127.95
RADIO AIDS TO NAVIGATION: NOTAM FILE RAP.
  RAPID CITY (H) VORTAC 112.3 RAP Chan 70 N43°58.56′
    W103°00.74′ 146° 61.3 NM to fld. 3160/13E.

Figure 78. *Airport/facility directory excerpt.*

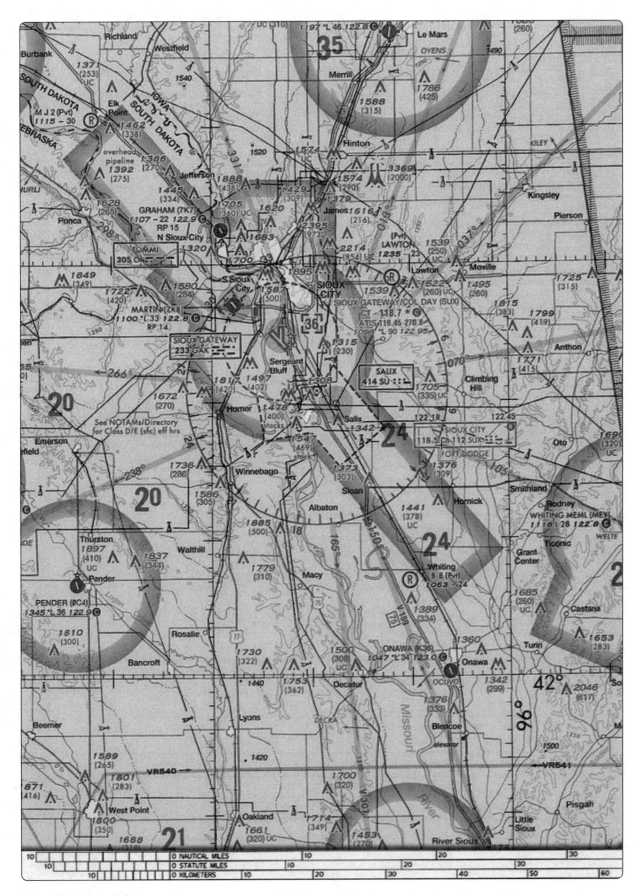

Figure 79. *Sectional chart excerpt.*

SIOUX CITY   N42°20.67′ W96°19.42′   NOTAM FILE SUX   OMAHA
  (L) VORTAC 116.5   SUX   Chan 112   313 ° 4.4 NM to Sioux Gateway/Col Bud Day Fld. 1087/9E.   HIWAS.   L–12I
      VOR unusable:
        280°–292° byd 25 NM   306°–350° byd 20 NM blo 3,000 ′
        293°–305° byd 20 NM blo 4,500 ′   350°–280° byd 30 NM blo 3,000 ′
        293°–305° byd 35 NM
      RCO 122.45 122.1R 116.5T (FORT DODGE RADIO)

SIOUX CITY

SIOUX GATEWAY/COL BUD DAY FLD   (SUX)   6 S   UTC –6(–5DT)   N42°24.16′ W96°23.06′   OMAHA
  1098   B   S4   FUEL 100LL, 115, JET A   OX 1, 2, 3, 4   Class I, ARFF Index—See Remarks   H–5C, L–12I
    NOTAM FILE SUX   IAP, AD
  RWY 13–31: H9002X150 (CONC-GRVD)   S–100, D–120, 2S–152,
    2D–220   HIRL
    RWY 13: MALS. VASI(V4L)—GA 3.0 ° TCH 49′. Tree.
    RWY 31: MALSR. VASI(V4L)—GA 3.0 ° TCH 50′.
  RWY 17–35: H6600X150 (ASPH-PFC)   S–65, D–80, 2S–102,
    2D–130   MIRL
    RWY 17: REIL. VASI(V4R)—GA 3.0 ° TCH 50′. Trees.
    RWY 35: PAPI(P4L)—GA 3.0 ° TCH 54′. Pole.
  LAND AND HOLD SHORT OPERATIONS

LANDING	HOLD SHORT POINT	DIST AVBL
RWY 13	17–35	5400
RWY 17	13–31	5650

  ARRESTING GEAR/SYSTEM
    RWY 13 ←BAK–14 BAK–12B(B) (1392′)
                        BAK–14 BAK–12B(B) (1492′) →RWY 31
  AIRPORT REMARKS: Attended continuously. PAEW 0330–1200Z ‡ during
    inclement weather Nov–Apr. AER 31–BAK–12/14 located (1492 ′)
    from thld. Airfield surface conditions not monitored by arpt
    management between 0600–1000Z ‡ daily. Rwy 13–BAK–12/14
    located (1392′) from thld. All A-gear avbl only during ANG flying ops. Twr has limited visibility southeast of
    ramp near ARFF bldg and northeast of Rwy 31 touchdown zone. Rwy 31 is calm wind rwy. Class I, ARFF Index
    B. ARFF Index E fire fighting equipment avbl on request. Twy F unlit, retro-reflective markers in place. Portions
    of Twy A SE of Twy B not visible by twr and is designated a non-movement area. Rwy 13–31 touchdown and
    rollout rwy visual range avbl. When twr clsd, ACTIVATE HIRL Rwy 13–31; MIRL Rwy 17–35; MALS Rwy 13;
    MALSR Rwy 31; and REIL Rwy 17—CTAF.
  WEATHER DATA SOURCES: ASOS (712) 255-6474.   HIWAS 116.5 SUX. LAWRS.
  COMMUNICATIONS: CTAF 118.7   ATIS 119.45   UNICOM 122.95
    SIOUX CITY RCO 122.45 122.1R 116.5T (FORT DODGE RADIO)
  ® SIOUX CITY APP/DEP CON 124.6 (1200–0330Z ‡)
  ® MINNEAPOLIS CENTER APP/DEP CON 124.1 (0330–1200Z ‡)
    SIOUX CITY TOWER 118.7 (1200–0330Z ‡)   GND CON 121.9
  AIRSPACE: CLASS D svc 1200–0330Z‡ other times CLASS E.
  RADIO AIDS TO NAVIGATION: NOTAM FILE SUX.
    SIOUX CITY (L) VORTAC 116.5   SUX   Chan 112   N42 °20.67′ W96°19.42′   313° 4.4 NM to fld. 1087/9E.
      HIWAS.
    NDB (MHW) 233   GAK   N42°24.49′ W96°23.16′   at fld.
    SALIX NDB (MHW/LOM) 414   SU   N42°19.65′ W96°17.43′   311° 6.1 NM to fld. Unmonitored.
    TOMMI NDB (MHW/LOM) 305   OI   N42°27.61′ W96°27.73′   128° 4.9 NM to fld. Unmonitored.
    ILS 109.3   I–SUX   Rwy 31   Class IT.   LOM SALIX NDB. ILS Unmonitored when twr clsd. Glide path
      unusable coupled approach (CPD) blo 1805 ′.
    ILS 111.3   I–OIQ   Rwy 13   LOM TOMMI NDB. Localizer shutdown when twr clsd.
    ASR (1200–0330Z‡)

SNORE   N43°13.96′ W95°19.66′   NOTAM FILE SPW.   OMAHA
  NDB (LOM) 394   SP   121° 6.8 NM to Spencer Muni.

SOUTHEAST IOWA RGNL   (See BURLINGTON)

Figure 80. *Airport/facility directory excerpt.*

Figure 81. *Sectional chart excerpt.*

CRAWFORD (99V) 2 W UTC −7(−6DT) N38°42.25′ W107°38.62′ **DENVER**
6470 S2 OX 4 TPA—7470(1000) NOTAM FILE DEN **L−9E**
RWY 07–25: H4900X20 (ASPH) LIRL (NSTD)
 RWY 07: VASI (NSTD). Trees. **RWY 25:** VASI (NSTD) Tank. Rgt tfc.
RWY E–W: 2500X125 (TURF)
 RWY E: Rgt tfc. **RWY W:** Trees.
AIRPORT REMARKS: Attended continuously. Rwy 07–25 west 1300 ′ only 25′ wide. Heavy glider ops at arpt. Land to the
 east tkf to the west winds permitting. 100LL fuel avbl for emergency use only. Pedestrians, motor vehicles, deer
 and wildlife on and invof arpt. Unlimited vehicle use on arpt. Rwy West has +15 ′ building 170′ from thld 30′ left,
 +10′ road 100′ from thld centerline. +45′ tree 100′ L of Rwy 07 extended centerline 414 ′ from rwy end. −8′ to
 −20′ terrain off both sides of first 674 ′ of Rwy 25 end. E–W rwy occasionally has 6 inch diameter irrigation
 pipes crossing rwy width in various places. Rwy 07 has 20 ′ trees and -10′ to 20′ terrain 20′ right of rwy first
 150′. E–W rwy consists of +12 inch alfalfa vegetation during various times of the year. Arpt lgts opr
 dusk–0800Z‡. Rwy 07 1 box VASI left side for local operators only or PPR call 970-921-7700 or
 970-921-3018. Rwy 07-25 LIRL on N side from Rwy 25 end W 3800 ′. Rwy 07 1300′ from end E 300′. No thld
 lgts Rwy 07-25 3800′ usable for ngt ops.
COMMUNICATIONS: CTAF/UNICOM 122.8
RADIO AIDS TO NAVIGATION: NOTAM FILE MTJ.
 MONTROSE (H) VORW/DME 117.1 MTJ Chan 118 N38 °30.39′ W107°53.96′ 033° 16.9 NM to fld. 5713/12E.

CREEDE

MINERAL CO MEM (C24) 2 E UTC −7(−6DT) N37°49.33′ W106°55.79′ **DENVER**
8680 NOTAM FILE DEN **H–3E, L–9E**
RWY 07–25: H6880X60 (ASPH) S-12.5, D-70, 2D-110
 RWY 07: Thld dsplcd 188 ′. **RWY 25:** Road.
AIRPORT REMARKS: Unattended. Elk and deer on and invof arpt. Glider and hang glider activity on and in vicinity of
 arpt. Mountains in all directions. Departure to NE avoid over flight of trailers and resident homes, climb to 200 ′
 above ground level on centerline extended prior to turn. Acft stay to right of valley on apch and/or departure
 route. 2′ cable fence around apron.
COMMUNICATIONS: CTAF 122.9
RADIO AIDS TO NAVIGATION: NOTAM FILE DEN.
 BLUE MESA (H) VORW/DME 114.9 HBU Chan 96 N38 °27.13′ W107°02.39′ 158° 38.1 NM to fld. 8730/14E.

CUCHARA VALLEY AT LA VETA (See LA VETA)

DEL NORTE

ASTRONAUT KENT ROMINGER (8V1) 3 N UTC −7(−6DT) N37°42.83′ W106°21.11′ **DENVER**
7949 NOTAM FILE DEN **H–3E, L–9E**
RWY 06–24: 6050X75 (ASPH) 1.1% up SW
RWY 03–21: 4670X60 (TURF-DIRT)
 RWY 21: Mountain.
AIRPORT REMARKS: Unattended. Wildlife on and invof arpt. Unlimited vehicle access on arpt. Mountainous terrain
 surrounds arpt in all directions.
COMMUNICATIONS: CTAF 122.9
RADIO AIDS TO NAVIGATION: NOTAM FILE ALS.
 ALAMOSA (H) VORTACW 113.9 ALS Chan 86 N37 °20.95′ W105°48.93′ 298° 33.7 NM to fld. 7535/13E.

Figure 82. *Airport/facility directory excerpt.*

Figure 83. *Altimeter.*